green guide

MAMMALS

OF AUSTRALIA

Terence Lindsey
Series Editor: Louise Egerton

Published in Australia by
New Holland Publishers (Australia) Pty Ltd
Sydney • Auckland • London • Cape Town
14 Aquatic Drive Frenchs Forest NSW 2086 Australia
218 Lake Road Northcote Auckland New Zealand
Garfield House 86 Edgware Road London W2 2EA United Kingdom
80 McKenzie Street Cape Town 8001 South Africa

First published in 1998 by New Holland Publishers (Australia) Pty Ltd
Reprinted in 2000, 2001

National Library of Australia Cataloguing-in-Publication Data:
Lindsey, Terence, 1941–.
Mammals of Australia
Includes index

ISBN 1 86436 307 X

1. Mammals—Australia I. Title.
(Series: Green guide (New Holland)).

599.0994

Publishing General Manager: Jane Hazell
Publisher: Averill Chase
Series Editor: Louise Egerton
Design and Cartography: Mark Seabrook
Picture Researcher: Bronwyn Rennex
Reproduction by: DNL Resources
Printed and bound in Singapore by Tien Wah Press

CONTENTS

An Introduction to Mammals 4

Egg-laying Mammals 8

Pouched Mammals 14

Placental Mammals 54

Introduced Mammals 72

Marine Mammals 82

A Checklist of Australian Mammals 92

Index 96

Further Reading, Wildlife Clubs and Organisations 97

An Introduction to Mammals

This guide is an introduction for beginners to the major groups of Australian mammals with an emphasis on those species the amateur naturalist and nature-lover is most likely to see.

Diversity

Of the world's approximately 4,200 species of mammals, about 350 have been recorded in Australia. Almost all are endemic. In round figures, about 60 species are rodents, about 70 species are bats, about 50 are whales and similar marine mammals (many of which barely touch Australian shores), and just under 20 animals are exotic — species deliberately introduced by humans and now self-sustaining in the wild. The remaining 150 are made up of the Echidna, the Platypus, and all the pouched animals, the marsupials. Among the marsupials are the kangaroos, wallabies, koala, wombats and a host of other uniquely Australian mammals.

Where They Live

Mammals are found in all habitats from deserts to rainforest and from the open ocean to alpine meadows. Many, such as the larger kangaroos and wallabies, are easy to see but others can be very difficult to observe or study because they are active only at night or are small, shy and retiring. Some are very rare or endangered. Quite often they are extremely difficult to identify, especially among the smaller carnivorous marsupials, the rodents and the bats.

A group of Eastern Grey Kangaroos. Kangaroos live mainly in open grassland dotted with trees.

The Common Wombat is widespread in grassy parts of eastern Australia.

Why Are Mammals Different?

Two mammalian features that set them apart from all other animals are, first, that their bodies are clad in fur, and second, that mammals suckle their young. Female mammals secrete milk that nurtures the infant until its own digestive system has matured sufficiently to enable it to be 'weaned' onto the adult diet. A third mammalian characteristic is less obvious: mammals chew their food. Chewing increases the speed and efficiency of digestion, and thereby contributes to the animal's ability to maintain a high metabolism. Different diets impose different stresses on teeth, which evolve accordingly, and so dentition (the form and arrangement of teeth) in mammals is extremely complex. Many small mammals are much more easily told apart by their teeth than by any other external feature.

Another striking characteristic of mammals is their range in size: of nearly 4,200 species of mammals on Earth, the largest, the Blue Whale, is several million times bigger than the smallest, which is probably a tiny bat.

WORDS TO KNOW

Talking about animals is easier if we know a few special words or recognise certain familiar words as being used in a special way.

Arboreal: tree dwelling.
Diurnal: normally active by day.
Endemic: native to a particular place and found nowhere else.
Feral: Introduced animals with self-sustaining populations.
Gestation: pregnancy.
Gregarious: gathered in a group, not necessarily interacting.
Home range: the sum of all the places that an animal visits in its normal daily routine.
Introduced: an animal that has been transplanted by humans and set free in the wild.
Migratory: normal movement from one part of the world to another.
Nocturnal: normally active at night.
Range: the sum of all the places that an animal species occurs.
Sedentary: remaining within a defined area.
Social: living in a group.
Solitary: normally living alone when adult.
Terrestrial: ground dwelling.
Territory: a home range defended against trespassers.

What Sort of Mammals are There?

Dingoes are placentals, a mammal group that also includes wolves, cats and lions.

Mammals come in three basic kinds or models, and Australia, together with New Guinea, has the unique distinction of being the only place in the world where all three can be found together. These three kinds of mammals are the placentals, the pouched mammals or marsupials and the egg-laying mammals or monotremes. They are distinguished by many features but perhaps most conspicuously by the different ways in which they reproduce.

The Placentals

Humans are placentals, as are dogs, cats, cows, horses, rats, mice and so on. These mammals give birth to young at an advanced stage of development, relying heavily on a sophisticated organ, the placenta, to nurture the baby while it remains within the mother's body.

In many cases, such as in horses, sheep and antelopes, the advancement is so marked that the infant can run almost immediately after birth.

The Marsupials

Australia happens to be full of mammals that do things differently. Marsupials, for

Marsupials, such as kangaroos, are distinguished from other mammals by rearing their young in a pouch.

example, give birth to their young when they are very tiny and the early stages of growth take place in a special pouch on the mother's abdomen. Australia is often thought of as the land of marsupials but there are other species in South America and one, the Virginia Opossum, is common across much of North America.

Koalas are marsupials that are active mainly at night.

The Monotremes

The third major group, the egg-laying mammals or monotremes, is made up of only three species: the Platypus and two echidnas. Australia and New Guinea are the only places in the world where these three mammals live. They are special in one striking respect: they lay eggs that hatch outside the mother's body, as birds do. Like all mammals, monotremes suckle their young, but there are no teats; the milk merely oozes from enlarged pores in the mother's skin.

The Platypus is a monotreme, a mammal group that lays eggs, as birds do.

Egg-Laying Mammals

Platypus

The Platypus is entirely aquatic, foraging for its food exclusively underwater.

The Platypus is in a group all of its own. Strictly aquatic, it inhabits the rivers and streams of eastern Australia. It is common in some areas but rare in others, partly because it needs entirely undisturbed, unpolluted waters. Males, which are substantially bigger than females, average about 50 cm in length and weigh up to 2 kg, but size varies markedly with locality.

Life in the Water

Platypuses live in tunnels dug into riverbanks. Periodically they emerge to forage. While underwater, the eyes, ears and nostrils are held shut; the forelimbs are used for braking and steering and the hindlimbs for propulsion. Platypuses feed on yabbies and other small aquatic animals, which they store in their cheek pouches and bring to the surface to eat. Adults lack teeth, instead they use horny plates along their jaws to chew. Especially in the evening, they float on the surface to groom themselves, and occasionally they haul out for a while onto some partly submerged log or boulder. Mostly solitary animals, their only call is a low growl, which is seldom heard by humans.

In spring, about two weeks after mating, the female lays 1–3, usually 2, eggs in a chamber at the end of a specially constructed nursery burrow. She incubates them by tucking them between her belly and her curled-up tail until they hatch in about 10–14 days.

HOAXED

The Platypus exhibits such a bizarre mix of features that when the first specimen reached Britain in 1798, it was widely regarded as a hoax. It was thought that some skilful taxidermist had cunningly sewn bits of various animals together to make the skin.

Echidnas

The Short-beaked Echidna (above) relies on its sharp spiny coat for protection. It will often roll into a ball (right) when alarmed.

There are only two species of echidna in the world. The Long-beaked Echidna inhabits the highlands of New Guinea, while the Short-beaked Echidna occurs in the lowlands and virtually throughout Australia.

The Short-beaked Echidna's identity is immediately obvious by its spines alone, although the fur of Tasmanian individuals is so long and dense it partly conceals them. This animal lives in almost all terrestrial environments from coastal forests to alpine meadows and interior deserts — in fact, it has the widest distribution of any native Australian mammal.

Habits

Adult Short-beaked Echidnas are about 50 cm long and weigh up to about 7 kg. They lack teeth but have a long sticky tongue with which they lap up ants and termites. To break into the nests they use their powerful forepaws. In some localities they appear to favour ants over termites, in others the reverse is true. Usually sedentary and solitary, they wander apparently at random over extensive home ranges that seem to lack defended boundaries or permanent dens of any kind. If disturbed in the open, they burrow straight down into all but the hardest ground using their claws and great heaves of their extraordinarily powerful bodies.

A STEEP GROWTH CURVE

Baby echidnas have among the steepest growth curves yet measured among mammals. At hatching, the tiny echidna weighs about 0.32 g. Within 45 days it is about 178 g – a 500-fold increase.

How does a Platypus Find its Food?

It has only recently been discovered that Platypuses locate their prey underwater using a sensory system previously unknown among mammals. This system exploits the fact that all animals generate a minute electrical pulse when they flex their muscles. The system was discovered almost by accident when it was found that captive Platypuses in a laboratory tank ignore dead batteries but investigate live ones. Much remains to be learnt about how exactly this system works but it has been demonstrated that when a freshwater shrimp — a favourite food of Platypuses — flicks its tail, it generates a disturbance in the surrounding electrical field that can be detected by a Platypus at a distance of about 10 cm. The electrical receptors are tiny pores that appear in rows on the soft rubbery skin of the Platypus's bill.

Platypuses feed on yabbies and similar small aquatic animals.

The Platypus detects its prey using a unique 'electrical' sensory system housed in its duck-like bill.

What are Male Platypus's Spurs for?

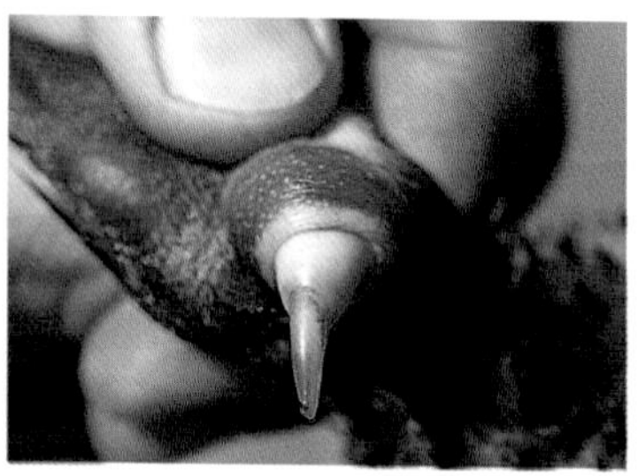

On each ankle the male Platypus carries a sharp, venomous spur of unknown function.

A puzzling oddity of Platypus anatomy is that males have a sharp, hollow spur on each ankle, which is connected to a venom gland in the groin. (Echidnas also possess a similar spur but there is no venom gland.)

The venom is so potent it is dangerous even to humans, but the mystery lies in what this equipment might be used for. Maybe it is used in conflicts with rival males either for mates or for territory but this has yet to be demonstrated.

Is Hibernation a Hi-tech Strategy for Survival?

A hibernating animal goes to sleep and allows its temperature to fall to that of its surroundings for weeks at a time, and in the Northern Hemisphere many animals hibernate to escape the rigours of winter. Once upon a time, this behaviour was regarded as 'primitive'; then it was discovered that, amongst other complex abilities, hibernating animals can rouse themselves without waiting for warmer weather to trigger them into revival. So then hibernation came to be regarded as a highly evolved state, carrying the logical implication that any animal that hibernates must be 'advanced'. With the astonishing discovery, in 1987, that echidnas in Australia's southeastern high country hibernate through winter, zoologists are perplexed as to how an apparently 'primitive' animal like an echidna can exhibit such hi-tech behaviour. To quote one researcher, 'Never underestimate an echidna!'

Research has revealed that echidnas in the snow country hibernate through winter, although some awaken to mate.

Are Echidnas Smart?

Echidnas exhibit some surprises in the brain-power department. Both monotremes and marsupials differ from placentals in that they have no corpus calloseum ('wiring' linking the two hemispheres of their brain), yet the brain hemispheres of echidnas are deeply convoluted and they have a large neocortex (the component where all the reasoning goes on). These features are strongly suggestive of complex brain power, and echidnas perform very well in laboratory tests designed to assess learning, memory and other advanced mental processes. In fact in some tests they have outperformed cats.

Not just an ugly face — echidnas perform extraordinarily well in the brain-power department.

POUCHED MAMMALS

Tasmanian Devil

Once widespread on the Australian mainland, the Tasmanian Devil is now confined to Tasmania.

With its bulky head, powerful jaws, jet black fur and variable white markings, the Tasmanian Devil is difficult to mistake for any other animal and indeed it is the only species in its group. About the size of a small dog, it is the largest of Australia's surviving meat-eating marsupials. Males generally weigh around 10 kg, females about 6 kg. Once widespread on the Australian mainland, it is now confined to Tasmania. It inhabits wooded country, from deep forests to the leafy outer suburbs of many towns and cities.

Nocturnal Hunting

Active mostly at night, Tasmanian Devils are essentially scavengers and carrion eaters, although they also catch birds, reptiles and small mammals when the opportunity arises and occasionally they raid henhouses. In fact devils will eat virtually anything of animal origin, including bones. They forage over home ranges that may be 10–20 sq km but these frequently overlap, so that it is not unusual for several individuals to be attracted to the same carcass. Such communal feasts are accompanied by a great deal of growling, screeching and scuffling; occasional bites may be serious but protracted fighting is rare. By day they sleep in underground burrows.

FOOTPRINTS

Tasmanian Devils leave distinctive tracks in snow, sand or mud. They are laid down in a sort of diamond pattern: one paw in front, then two side by side, and one at the back. Their normal gait is a slow awkward lope, but at speed they use an odd canter that has been described as 'reminiscent of a rocking-horse'.

Quolls

Roughly cat size, quolls are carnivorous marsupials that are unusual among Australian mammals in being conspicuously spotted with white. They rely heavily on insects and carrion for food but they catch many ground-nesting birds and small mammals, as well as eating fruit, grass and other vegetation. They range over wooded habitats and sometimes fossick around forest camp-grounds and picnic areas. Although mainly active at night, they occasionally forage during the day.

The Spotted-tailed Quoll (above) differs from its relative the Eastern Quoll (left) most obviously in the white spots on its long tail.

Distribution

The Northern Quoll is confined to tropical Australia and the Western Quoll, once widespread in the interior, is now common only in the far southwest. The Eastern Quoll and the Spotted-tailed Quoll overlap in much of their distribution. The former was once widespread in the south-eastern States but it suffered a severe decline early in the twentieth century — perhaps an epidemic of some unknown disease — from which it never recovered. It is now common only in Tasmania.

Distinguishing Features

Among Eastern Quolls, there are two colour forms, or morphs, which often occur together: one black, the other greyish fawn. Both have large, irregular white spots liberally scattered across the upperparts, but their tails are unmarked. This feature helps distinguish them from the larger, chocolate brown Spotted-tailed Quoll, in which, as the name indicates, the spots extend also onto the tail.

The Telltale Tail-tip

The Mulgara and the Kowari are very similar-looking carnivorous marsupials and both are about 30 cm long. While the Mulgara's fur is brownish rather than greyish, it is in their tails that the major difference lies. The Mulgara features a distinctive 'crest' of long black hair on the upper surface only of the tip of its tail. In the Kowari the long black fur completely encircles the tail to form a bushy tip. Also, while the Kowari is confined to stony gibber desert, the Mulgara inhabits sandy deserts in the far west.

The Mulgara's black tail-tip is bushy only on the upper surface.

The Mulgara and Kowari are members of a desert-dwelling trio that share similiarities in appearances and habits. The third member is the Kaluta, which is about half the size of the other two and is confined to the Pilbara in Western Australia. All three of these ground-dwelling predators feed on large insects, spiders and similar animals. Other ground-dwelling, desert-adapted carnivorous marsupials include the several species of mostly smaller dibblers and pseudoantechinuses.

The Kowari resembles the Mulgara except in its much bushier tail.

POUCHLESS

The Kowari is one of the few marsupials that lacks a fully-formed pouch. Instead, two loose folds of skin form on the mother's abdomen when her litter is born about 30 to 35 days after she has mated. Within these folds she has six, sometimes seven teats from which the young suckle for about 16 weeks.

Which Carnivorous Marsupials Live in Trees?

The two species of phascogales, the Red-tailed Phascogale and the Brush-tailed Phascogale, differ from all the other insect- and flesh-eating marsupials in that they live mainly in trees. Both have bushy tails, unlike other similar-looking small insect-eaters, such as the dibblers, pseudoantechinuses, ningauis and antechinuses. The Brush-tailed Phascogale is widespread in dry eucalypt woodland in much of coastal and near-coastal Australia but the Red-tailed Phascogale is confined to the far southwest.

The Red-tailed Phascogale lives mainly in trees.

Tiny but Mighty Wrestlers

The tiny ground-dwelling ningauis are ferocious and intrepid hunters. They often wrestle with large spiders and centipedes before despatching them with their needle-sharp teeth. They have grey-brown, rather shaggy fur with long, slender, sparsely furred tails. They are active at night and sleep by day deep inside a spinifex tussock or similar shelter. The largest of the three ningauis, the Wongai Ningaui, is about 13 cm from its nose to the tip of its tail. It is also by far the most widespread, its range extending into the interior parts of all mainland States except Victoria. The Pilbara Ningaui is confined to the Pilbara region and the Southern Ningaui to parts of southwestern New South Wales, the Nullarbor Plain and the region around Kalgoorlie.

The tiny but feisty ningaui often tackles prey nearly its own size.

Antechinuses

The Yellow-footed Antechinus is one of the most widespread of its group.

As a group, the antechinuses greatly resemble dibblers and several other small dasyurids except that they inhabit forests of one kind or another in coastal and near-coastal eastern and northern Australia. Generally somewhat bigger than a House Mouse, they live mostly on the ground and feed mainly on insects.

There are about eight species of antechinuses. With the exception of the Yellow-footed Antechinus, most species are very difficult to identify, especially in the heavily populated southeast where three dark brown species — the Brown, Dusky and Swamp Antechinuses — inhabit the wettest, densest forest areas.

SEX RATIOS

Among mammal litters, populations of males and females are usually about equal. In Brown Antechinus litters, however, the sexes are represented unequally with an average of 58 per cent being female.

Yellow-footed Antechinus

The Yellow-footed Antechinus is perhaps the most familiar member of the group because it is widespread in eastern Australia and also in the far southwest. It prefers the more open, drier forests and woodlands of the southeast and is common in suburban gardens, even entering houses to fossick in kitchen cupboards for food. It is readily identified by its black tail-tip and the marked shift in fur colour from its greyish head to its yellowish red rump.

Dunnarts

Big ears and a stout tail characterise the Fat-tailed Dunnart (above); similar species include the Long-tailed Dunnart (left).

There are about 19 species of dunnarts. They are distributed across the Australian continent in such a way that few areas lack at least one species and some have several. Dunnarts vary a little in size but typically they are about 15–20 cm long, with greyish fur, pointed noses and beady black eyes. Their similar appearances make them difficult to tell apart but it is sometimes possible to work out a tentative identification on the grounds of locality alone: the White-footed Dunnart, for example, is the only species in Tasmania (although it also occurs on the nearby mainland). Over much of the arid interior, however, the situation is much more intricate because several species overlap widely in distribution, and may even occur in the same habitats.

The dunnart group also includes the Kultarr, which is very closely related despite its strikingly long hindlegs that give it the appearance more of a hopping-mouse than a dunnart.

Fat-tailed Dunnart

The Fat-tailed Dunnart is readily identified from other dunnarts by its unusually large ears and its conspicuously thickened tail. In times of plenty, surplus food is converted into fat and stored in the tail to be reabsorbed in leaner times, so the diameter of the tail is in effect a crude indicator of the prevailing food supply.

Who are the Marsupial Predators?

Marsupial predators are known as dasyurids. The Tasmanian Devil and the several species of quolls are by far the largest of the dasyurids, but the group is very diverse, and includes about 30 species of small insect-eaters, some not a lot bigger than a mouse. Despite their diminutive size, they are typically ferocious and frequently tackle large spiders and similar prey nearly as big as themselves.

The Tasmanian Devil is by far the largest of Australia's surviving carnivorous marsupials.

Although several small dasyurids have striking anatomical features — the planigales, for example, have extraordinarily flattened skulls that in some species are only about 6 mm deep — in general they differ very little in appearance. Distinguishing between them is an intricate and highly technical task but most of them can be sorted into one of two clusters according to their habitat and distribution. The planigales, antechinuses and phascogales are mostly woodland inhabitants of the east and north, while the dunnarts, pseudantechinuses, dibblers, nigauis and a handful of larger members that includes the Kultarr, Kowari and Kaluta are mainly desert and grassland inhabitants of the western interior.

The Brown Antechinus is little bigger than a mouse.

Both the Numbat and the rarely encountered Marsupial Mole (which spends almost its entire life underground) are also marsupial carnivores, but they appear to be only distantly related to the dasyurid family.

Sex and Ulcers: is there a Connection?

In the case of certain antechinuses and phascogales, the answer is yes, if they are male. Several of these species have remarkably brief breeding seasons, with females receptive to males for only about two weeks a year. Males live less than a year, so they have only one opportunity to breed. These small marsupials pursue the theoretical optimum strategy in such circumstances, which is to abandon all other activities in favour of mating as long as females are receptive. Captive individuals have been recorded copulating for 12 hours straight. Resulting stress levels are so high that all males fall victim to ulcers or similar disorders and die within a few days of mating.

Only a few weeks old, a brood of baby Dusky Antechinuses huddle in their mother's open pouch. Stress-induced ulcers mean their father will have died within days of mating.

Is the Tasmanian Tiger Extinct?

Although often called the Tasmanian Tiger because of its striped hindquarters, in some respects this magnificent animal more closely resembled a wolf. More correctly called the Thylacine, it was the largest of the recent dasyurids. Once widespread in Australia and New Guinea it apparently retreated before the spread of the Dingo several thousand years ago, and became restricted to Tasmania, probably within the last 2,000 years. With European settlement in Tasmania the Thylacine's diet broadened to include chickens and sheep. A bounty placed upon its head in 1830 lead to its rapid decline. It now seems almost certainly extinct, although unverified reports persist. The last certain Thylacine was an individual in Hobart Zoo that was captured in 1933 and died in 1936.

The last definite Thylacine, which died in 1936.

Bandicoots

Northern Brown Bandicoots (above) are widespread in eastern Australia. Right: An Eastern Barred Bandicoot.

Bandicoots of one species or another occur across Australia — or rather once occurred, because several arid interior species are extinct or very nearly so. The best known species is perhaps the Northern Brown Bandicoot, which has a mainly coastal and near-coastal distribution from the Hawkesbury River north to Cape York and west to the Kimberley (see map). It is the largest species, with males — which are a little bigger than females — reaching about 70 cm in total length and weighing 2 kg. It is common in forested environments, and often visits suburban gardens, fossicking for food.

The Long-nosed Bandicoot also often visits eastern suburban gardens but it is substantially smaller, with very much longer and more pointed ears and snout. The Eastern Barred Bandicoot remains fairly common in parts of Tasmania but has an extremely fragmented distribution on the mainland. A distant relative, the Rufous Spiny Bandicoot, inhabits the remote rainforests of Cape York.

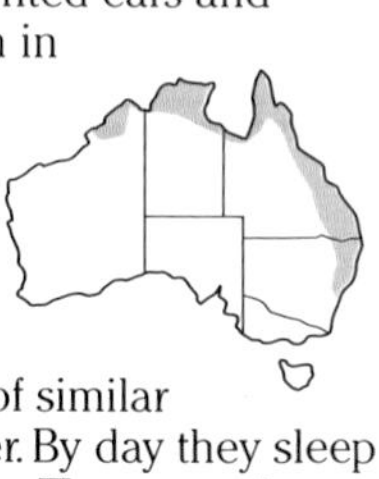

Solitary Habits

Bandicoots are solitary animals that forage on the ground at night for insects, spiders, seeds, berries and a wide range of similar fare, often digging into the soil or rummaging in the leaf litter. By day they sleep in well-concealed 'nests' that lack an obvious exit or entrance. These consist mainly of loose heaps of leaf litter scraped together on the ground, sometimes roughly roofed over with soil.

Bilbies

The Bilby is easily identified by its big 'rabbit' ears and distinctive white tail tip.

Until quite recently there were two bilbies on the Australian continent but the Lesser Bilby has not been reliably sighted since the 1960s and it appears to be extinct. With its soft grey-blue fur, long pointed snout, enormous ears and long half black, half white tail, the remaining species, known simply as the Bilby, is virtually unmistakable — although it is not easy to observe because it is rigidly nocturnal. It is a vigorous burrower, despite its deceptively small and delicate feet, and by day it sleeps in a chamber at the end of a burrow, sometimes several metres long and often with its entrance against a termite mound.

Size, Food and Distribution

Male Bilbies may be up 90 cm in length and weigh over 2 kg but females are substantially smaller. The species once ranged almost everywhere across the arid interior of Australia but it is now more or less confined to the far northwest, roughly from the Tanami Desert to Broome. Spinifex is usually abundant wherever it occurs. Digging for much of its food, it eats seeds, roots and insects, and — like many desert animals — can extract most of its water requirements from the food it eats. It lives usually alone or in pairs.

PREGNANCY

Among mammals, pregnancy is known as gestation. Bandicoots have the shortest known gestation periods of any mammal — barely over 12 days in a couple of species.

Koala

Virtually a cultural icon of Australia, the Koala is common in much of eastern Australia. It lives almost its entire life in trees but often descends to the ground to move between trees.

There is only one Koala species and it is unmistakable. The only tree-dwelling marsupial to lack a tail, the Koala has woolly fur and large fluffy ears. Its size varies widely according to age, sex and locality, but is typically about 50–60 cm long and about 5–10 kg in weight. Males are much bigger than females. Koalas are mostly solitary; although they spend much of their time in trees, they often come to the ground, where they can run nimbly and even swim.

Their distribution extends across much of eastern and southeastern Australia, but it is curiously patchy — Koalas are absent from some areas for no obvious reason, and their precise habitat requirements are still not entirely understood. Part of the restriction is that Koalas eat only the leaves of just a few species of gum (eucalypt) trees.

Habits and Home Range

Most feeding takes place just after dusk. Although they tend to occupy home ranges, usually around 3–4 ha in extent, walkabouts of 30 km or more have been recorded. Koalas sleep during the day. In cold weather they often select a large fork to huddle in, but in hot weather they may sprawl along the limb of a tree with their legs dangling.

Numbat

A banded back and 'bandit' mask render the rare Numbat almost impossible to misidentify.

The Numbat has no close relatives. About the size of a small cat, it is one of the most distinctive of Australian marsupials, with its dark 'bandit' mask and barred back. The sexes are alike but females are a little smaller than males.

Usually solitary, it forages during the day and sleeps in a hollow log or some similar shelter at night although — especially in winter — it may dig a burrow in the ground in which to sleep. While it eats only termites, it seldom attacks large earthen nests, instead using its sense of smell to locate nests in rotten logs or the concealed corridors that termites commonly construct between their nests and feeding areas. It breaks into these with scrabbling motions of its nimble forepaws, and extracts the insects from their galleries with rapid flicks of its long, sticky tongue.

Habitat and Range

Once widespread across arid southern Australia, it is now confined to a few scattered localities in the far southwest of the continent, where its existence is precarious. Its current habitat is eucalypt woodland, especially wandoo and jarrah, but it once also occurred in mulga, mallee and other kinds of scrub.

ALIEN THREAT

The introduced fox may be the single greatest serious threat to the Numbat's survival. In one monitored population, vigorous control of foxes in an area resulted in a 40-fold increase in Numbat numbers.

Why are Koalas so Dozy?

It's all to do with their diet of eucalypt leaves. These carry a formidable load of poisons as a defence against insects. They are also made up mostly of cellulose, a fibrous material that animals cannot digest. Koalas must therefore rely on bacteria living in their stomachs to do the job for them. The process is efficient but slow, which implies a hefty cargo of foliage in the stomach while it happens. The extra weight does not matter much to a cow chewing cud in a meadow but it does matter to an animal that must also be small enough to manoeuvre in the tops of trees.

The Koala's evolutionary response to this impasse has been to make do with less food. Its metabolism is only about half that of comparable mammals, and it also has a low 'activity budget' — Koalas spend about 80 per cent of a typical day asleep.

Koalas spend much of their lives asleep.

Why are Koalas Finicky Eaters?

A Koala's liver has an impressive armoury of chemical devices to neutralise the complex broth of eucalypt toxins that enter its bloodstream but this chemical wizardry consumes energy. The overall toxin load in eucalypt foliage varies enormously with age, season and many other factors. Only by careful selection of young leaves with low toxin levels can the Koala hope to derive an overall energy surplus, so they have become very finicky eaters.

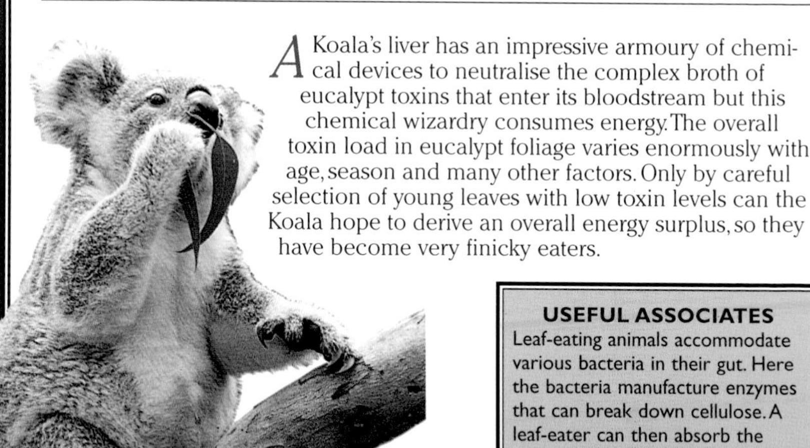

Koalas carefully select young leaves that are low in toxins.

USEFUL ASSOCIATES

Leaf-eating animals accommodate various bacteria in their gut. Here the bacteria manufacture enzymes that can break down cellulose. A leaf-eater can then absorb the resulting nutrients into its bloodstream. An association of this kind between two living creatures is known as symbiosis.

Which Large Australian Mammal Leads a Subterranean Lifestyle?

Among the largest of all mammals living mainly underground are the wombats. They dig elaborate warrens or systems of tunnels, which may extend as far as 20 m, with connecting passageways, multiple sleeping chambers and several entrances.

A Northern Hairy-nosed Wombat.

A favourite site for a warren is a grassy slope overlooking a stream or gully in open woodland. The big burly wombat, about 1 m long and weighing in at around 25 kg, lives alone. Each animal usually has several burrows within its home range of several hectares — but wombat home ranges often overlap.

Wombats emerge at dusk to feed on grass. Wandering as they graze, they may visit several burrows during the night and the one they select to retire to at dawn is not always the one they left at dusk; it may even be on a neighbouring wombat's patch.

The Common Wombat is widespread across much of southeastern Australia. Two other wombat species are very rare: the Southern Hairy-nosed Wombat is more or less confined to the Nullarbor Plain, and the Northern Hairy-nosed Wombat is now known only from the vicinity of Epping Forest National Park in central Queensland. These two species, unlike the Common Wombat, have pointed ears, silky fur and hairy noses.

The Common Wombat is both common and widespread, but its only two close relatives are very rare.

A Green Ringtailed Possum with distinctive eye patches.

Green Ringtailed Possum

Largely confined to the rainforests of the Atherton Tableland in far northern Queensland, the Green Ringtail is readily identified by its unique greenish fur. A small white patch below each eye is also distinctive. Adults have a total length of about 62 cm long and weigh about 1.1 kg. The Green Ringtail usually sleeps during the day in an exposed site on a horizontal limb but, unlike all other Australian ringtails, it is not rigidly nocturnal and on occasions it forages in daylight, especially on cloudy days. It feeds mostly on leaves. Babies sometimes utter a soft hissing sound but adults, which are usually solitary, seem to be entirely silent.

In the same forests as the Green Ringtails live Herbert River Ringtails, which are blackish above and white below, and Lemuroid Ringtails, distinguished by their long, greyish, woolly fur.

The Striped Possum is extremely agile.

Striped Possum

The bold, skunk-like black-and-white stripes of the Striped Possum's fur are unlike any other Australian mammal. Inhabiting New Guinea and the highland rainforests of northeastern Queensland, adult Striped Possums are about 26 cm long, with a bushy black 34 cm long tail. Active only at night, they are much more lively in the trees than ringtails but they are very shy; normally allowing no more than a glimpse before scampering away from the spotlight's beam. At times they may be quite noisy, uttering a range of growls, gruff shrieks and rasping notes.

By day they sleep in tree hollows. Although mainly solitary, pairs are sometimes seen together and occasionally share dens. Striped Possums feed on the larvae of wood-boring beetles and similar prey. By tapping the wood with their claws, they are presumably trying to find grubs; once located, they gnaw and gouge the surrounding timber apart to expose them, and use their elongated fourth finger to pick them from the cavity.

The very rare Leadbeater's Possum.

Leadbeater's Possum

A Leadbeater's Possum could be mistaken for a Sugar Glider except that it lacks a gliding membrane and its bushy tail widens rather than tapers towards the tip. So rare and little known that it was considered extinct until rediscovered in 1961, Leadbeater's Possum remains one of the most seriously endangered of Australian possums. It lives within a tract of eucalypt forest only about 1,000 sq km in extent in eastern Victoria and shows a marked dependence on large, mature mountain ash trees. It eats insects and various plant fluids, such as sap, honeydew, manna and nectar.

Rigidly nocturnal, Leadbeater's Possum lives in colonies of eightor more, all of which share a nest built of shredded bark in a tree cavity usually 10 m or more above the ground. Despite the lack of a gliding membrane, these shy, agile and very active creatures often make dramatic leaps between branches.

GLAND-LESS

One of the numerous, but less obvious, ways in which Leadbeater's Possum differs from other possums is in its lack of scent glands. Instead it apparently relies upon the odour of its saliva as a social signalling device.

The Sugar Glider can glide 50 metres or more.

Sugar Glider

Measuring about 30 cm from nose to tail-tip, the Sugar Glider may be easily confused with the Squirrel Glider, which sometimes occurs in the same area but tends to favour drier, more open woodlands away from the coast. A little smaller than the Squirrel Glider and about half its weight, the Sugar Glider also has a much less bushy tail and a wider head. Sugar Gliders live in pairs or small parties in eucalypt woodlands almost throughout northern and eastern Australia, as well as in New Guinea. They feed mainly on the gum and sap of eucalypts, which they lap up from grooves chiselled in the bark of trees with their sharp teeth but they also eat fruit, nectar and pollen as well as insects and spiders. They are active at night and sleep in tree hollows during the day.

Greater Gliders are active at night.

Greater Glider

The Greater Glider's large size, big fuzzy ears and long woolly coat are distinctive. About half of its 1 m total length consists of a fluffy, partly prehensile tail. Its colour ranges from black through brownish grey to nearly white. The Greater Glider is capable of breathtaking glides between trees and it can even turn through 90 degrees. Unlike other gliders, the gliding membrane extends only to the elbow, not to the wrist, and it has the distinctive habit of tucking its forepaws under its chin as it glides.

CREATURES OF HABIT
Greater Gliders tend to use the same pathways night after night as they forage. Their presence can sometimes be detected by the fine, horizontal scratch marks they leave on tree trunks at the points of take-off and landing.

Greater Gliders are silent, mainly solitary and strictly nocturnal. They spend the day in cavities high in the tall trees of most kinds of wooded areas, although they avoid rainforest. They are heavily dependent upon tall old trees with roomy hollows, which means they are extremely sensitive to unsympathetic logging practices and are often only truly common in undisturbed old-growth forests.

The distinctive Spotted Cuscus.

Spotted Cuscus

With its tubby build, inconspicuous ears and striking white and brown spotted coat, the Spotted Cuscus is unmistakable. Males may reach a total length of almost 1 m and weigh nearly 6 kg; females are a little smaller. The Spotted Cuscus prefers forests along rivers and streams. Although in Australia it is confined to Cape York Peninsula, it is widespread in New Guinea, where it may occur in lightly wooded grasslands and even around villages. It is often seen during the day because, like the Green Ringtail, it sleeps in the open on a large limb, usually curled up with its head between its knees. At night it forages in the canopy, eating mainly leaves. There are many cuscus species in Indonesia and New Guinea but this and the Southern Common Cuscus, which is greyish brown and very much smaller, are the only two Australian species.

A Honey-possum seeking nectar.

Honey-possum

In many ways the Honey-possum is one of the oddest Australian mammals and it is unique in several aspects of its feeding and breeding behaviour. At just under 20 cm long, a little over half of which is tail, and weighing around 10 g, it is a tiny possum and its long, slender snout and dark band of fur down the spine are distinctive. Females tend to be heavier than males and are dominant. Honey-possums inhabit far southwestern Australia, approximately from Geraldton to Esperance. Mainly sedentary, their chief habitat is sandy heath with abundant flowering plants, especially those in the families Proteaceae and Myrtaceae; here they are often very common. They do not appear to be under any immediate threat and in conservation terms they are listed as 'secure'.

VIRILITY TO FLAUNT

The testes of male Honey-possums are enormous, making up about 4 per cent of their total body weight. They are so large that they form a sort of cushion on which the animal sits when it assumes an upright posture. Male Honey-possums also have the longest sperm of any mammal, including whales.

This species of possum is confined to alpine regions.

Mountain Pygmy-possum

The Mountain Pygmy-possum is among the most endangered of all possums. All known colonies lie within about 30 km of Mount Hotham in Victoria and Mount Kosciuszko in New South Wales and it occurs only in areas of boulder scree among snow gums and mountain plum pine at about 1,600 m and above. All the pygmy-possums are minute but this is the largest, with a total length of about 25 cm, more than half of which is a prehensile, largely naked tail. In summer it preys heavily on Bogong Moths but in winter it relies on seeds which it has previously stored. In very cold weather it often sleeps for weeks at a time.

Of the other pygmy-possums in Australia, the Long-tailed Pygmy-possum lives in the northeast, the Western Pygmy-possum in the southwest and the Little and Eastern Pygmy-possums in the east and southeast, including Tasmania.

How to observe Tree-dwelling Mammals

Spotlighting at night with a powerful torch is a good way to observe nocturnal mammals.

A good way of observing possums and other tree-dwelling mammals is with the use of a spotlight or powerful torch. The trick is to hold the torch as close to your eyes as possible so that you are looking directly down the beam. The reason for this is that many nocturnal animals have a structure called the tapetum inside the eye, which acts as a mirror, so the light from your torch is 'bounced' right back at you, much dimmer but along the same axis. The closer your own eye is to this axis, the more likely you are to detect the reflected light, which returns as 'eyeshine'. With practice and a suitably powerful light, you can even use binoculars quite successfully. Remember to use your ears as well as eyes — Striped Possums, for example, make a lot of noise crashing about in the foliage and are much more often heard than seen.

Most possums are active only at night.

BUSH COURTESY

Do not hold your animal under observation for too long. From a possum's perspective, a good nightly dazzling is no help at all in finding supper.

How does an Animal Fall Slowly?

Mobility is a serious problem for many tree-dwelling animals that cannot fly. Sooner or later, a tree's resources, whether it be fruit, leaves, pollen or sap, are used up and animals must move on. In dense forest, where trees are crowded and branches interlock, an animal can scramble or even leap from one tree to another but when trees grow more scattered, as in eucalypt woodland, there is no alternative but to climb down one tree and laboriously scale another. Gliding membranes may have evolved as a device for reducing such energy expenditure — at best it may be possible to reach the trunk of a neighbouring tree by gliding; at worst it is much less effort and probably far less dangerous to glide rather than scramble down to the ground. Either way it is an improvement and it seems to have been so effective among Australian possums that the device has evolved at least three different times in three different animals: in the Greater Glider, the Sugar Glider and the Feathertail Glider.

The Squirrel Glider is an accomplished glider.

Little Pygmy-possums: members of our diverse possum lineage.

POSSUM EVOLUTION

The possums probably evolved in Australia: 25 of a total of about 60, nearly half, now live here. All possums are at least partly arboreal, although the Rock Ringtail of Arnhem Land lives among rocks and climbs trees only to forage at night. Most are nocturnal and solitary but a few, such as the Sugar Glider and Leadbeater's Possum, live in social groups.

Feathertail Glider

The Feathertail Glider is one of the smallest of the gliding mammals.

No other animal can be compared with this species. It is instantly recognisable by its minute size — less than 15 cm long and roughly 10 g in weight, as well as its gliding membrane and its unique 'feathered' tail. Although widespread in eastern Australia, from South Australia to Cape York, it is absent from Tasmania. It inhabits wooded areas from tall, dense eucalypt forests to open woodlands, even in city suburbs. Strictly nocturnal, it often forages for nectar, sap, flowers, leaves and small insects in shrubbery, where it is vulnerable to night-prowling cats.

REMARKABLE GRIP

The Feathertail Glider is unusual in having an extra pad on the sole of its foot which, together with other details of its foot structure, allow it to scamper over vertical surfaces — even panes of glass.

Family Life

Feathertail Gliders live in family groups typically made up of a mated pair with one or two litters of young. They live in spherical nests built of overlapping leaves, usually in the cavity of eucalypt trees but sites vary: they have been found, for example, snuggled up in the plastic bags used to cover bananas. A dozen or more individuals may occupy a single nest but three or four is more usual. Infants are carried in the pouch for about their first 60 days after which their weight begins to impede their mother's gliding abilities so they are left in the nest.

Brushtail Possums

The Common Brushtail Possum is a familiar nightly garden visitor. Right: a Mountain Brushtail Possum.

Brushtail possums are approximately the size of a small cat with males being much larger than females. These nocturnal creatures are easily recognised by their dark, bushy tails, which are only slightly prehensile. They have varied diets and often utter noisy guttural coughs and hisses, especially during the breeding season. The Common Brushtail Possum is the most familiar possum in the southeast (see map). It is common in wooded areas, even in the parks of major cities. In suburban gardens it often fossicks in litter bins and raids bird feeders. It has prominent ears and its fur is variable in colour, usually brownish grey.

Mountain Brushtail Possum

Easily confused with its close relation, the Common Brushtail Possum, the distributions of these two species sometimes overlap. The Mountain Brushtail, however, is normally confined to the dense, wet sclerophyll forests of southeastern Australia and its fur is generally thicker. Both have prominent ears but those of the Common Brushtail are larger.

DISTANT RELATIVE

Brushtails are familiar woodland animals in the heavily populated southeast, but they have at least one close relative in a very different habitat in a far away place. The little-known Scaly-tailed Possum occurs only in the remote Kimberley, where it lives among rocks. Apart from its range and habitat, its most distinctive feature is its unique, coarsely-scaled tail.

Which Possum has become a Pest?

Beginning in 1858, several consignments of Brushtail Possums were released in New Zealand with a view to establishing a fur industry. The Brushtails are now common throughout the main islands and reach even remote Fiordland. They have proliferated with such vigour that their densities in some areas are now six times greater than those in Australian forests. The species has become a major pest, damaging vast areas of native forest.

How do Possums Communicate?

A sophisticated sense of smell is a distinctive mammalian characteristic and many groups use scent to communicate between individuals of the same species. Australian possums have scent glands on the chest or elsewhere, depending on species. These glands exude a special smelly substance that they rub onto the bark of trees as a way of 'sign-posting' their territorial boundaries or other socially important pieces of information: such is possum-speak.

Mother Ringtail Possums commonly tote their babies as they hunt.

What is a Prehensile Tail?

Many long-tailed mammals use their tails as a sort of fifth limb. In some the musculature of the tail is sufficiently strong, intricate and well-cordinated to enable the animal to support its entire dangling weight by curling the tip around a tree branch as some monkeys do. Such tails are described as 'prehensile'.

A ringtail's tail can support its entire body weight.

Which Australian Marsupial lives in a Drey?

In Australia, many tree-living mammals rely on natural cavities in trees for shelter. Natural cavities are safe and snug but the supply is limited and they tend to be common only in very old forests. Competition from other animals needing shelter is keen and cavities need to be defended. A few Australian mammals have overcome these difficulties by building their own shelters. The Common Ringtail Possum, for example, sleeps through the day in a specially built nest of sticks called a drey. Each animal normally has several dreys and often several neighbouring individuals build dreys close together. This roughly cat-sized possum is common in wooded habitats, including urban parks and gardens, across much of southern and eastern Australia. Contrasting with its grey to brown fur are its white ear tufts and the white tip to its long, tapered, strongly prehensile tail. Like other possums it is active at night, especially from dusk to midnight, when it feeds chiefly on leaves.

Ringtail families often live together in their drey until the youngsters mature and disperse, and the pair bond may persist for several years. Twins make up the usual litter (although the mother has four teats), and they are weaned at about six months of age.

POO-EATERS

The Common Ringtail is a coprophore, that is it eats its own faeces but only the soft faeces it produces by day when it dozes in its nest; from these it extracts various proteins and B-group vitamins. The dry hard faeces the ringtail produces at night are not eaten.

The Common Ringtail Possum frequently builds its own shelter, called a drey.

Bettongs

The Tasmanian Bettong is now confined to Tasmania.

Bettongs are medium-sized ground-dwelling marsupials with powerful forefeet for digging and a tail prehensile enough to carry nesting material. There are five species.

The Burrowing Bettong and the Brush-tailed Bettong are endangered and now largely confined to the far southwest, and the Northern Bettong lives only in a few localities on the Atherton Tableland in far northeastern Queensland. The Rufous Bettong remains widespread in eastern Australia, although rare and scattered (see map). The Tasmanian Bettong was once also widespread in southeastern Australia but it became extinct on the mainland probably sometime before 1900; it remains common in Tasmania, where it is most numerous in dry forests of peppermint gums and silver wattles with a grassy understorey.

Tasmanian Bettong

About 60 cm in total length, the Tasmanian Bettong might be confused with the Long-nosed Potoroo, which sometimes occurs in the same forests but normally requires denser undergrowth. It is, however, slightly smaller and paler. It is strictly nocturnal and spends the day in a nest of grass and bark fragments which it constructs under a fallen log or in a similarly secluded place. It eats a wide range of roots, shoots, seeds and underground fungi.

Potoroos

Potoroos can be distinguished from bettongs most easily by their shorter tails.

Potoroos resemble bettongs, both in size and lifestyle, but the potoroos have a more tapered head and pointed snout, a less robust build and a shorter tail. They also differ in their habitat, preferring to live in dense heaths or tall, wet forests and rainforests rather than in dry forests.

Also like the bettongs (and the bandicoots), the potoroos dig for much of their food, scrabbling with their sharp-clawed forepaws to unearth roots, tubers, fungi and subterranean vegetation, as well as insects and spiders. They are solitary animals and strongly nocturnal, spending the day in roughly built well-concealed nests of grass and bark lining a shallow scrape in the ground.

RAINFOREST RELATIVE
Although in the same family as the bettongs and potoroos, the Musky Rat-kangaroo is active by day. Confined to the rainforests of far northern Queensland, it moves around in the gloom of the forest floor like a rabbit. Often venturing into shrubs and the lower branches of trees, it is not especially shy or hard to find.

Long-nosed Potoroo

Locally common in eastern and southern Australia, including Tasmania, the Long-nosed Potoroo (see map) is about 60 cm in total length but its size and weight vary greatly with locality. It is easily confused with the Long-footed Potoroo, which is, however, a little larger and confined to east Gippsland and a few neighbouring localities in Victoria.

Hare-wallabies

The Spectacled Hare-wallaby is the only widespread member of its group.

This is a group of wallabies that owe their name to a fancied resemblance to the European Hare: they live in open country, are solitary in habits, spend the day motionless in a 'hide' or 'set' and bolt from cover with prodigious bursts of speed when alarmed. Of the four species, the Eastern Hare-wallaby was last reported in 1890 and is known from only a handful of museum specimens, the Central Hare-wallaby is known from only a single skull and the reports of desert Aborigines, and the last known wild mainland population of the Rufous Hare-wallaby was wiped out by wildfire in 1991. Fortunately, the Rufous Hare-wallaby still survives on several islands in Shark Bay, Western Australia, and there are thriving captive colonies.

Spectacled Hare-wallaby

The Spectacled Hare-wallaby is the chief survivor of this group. It still lives in hummock grasslands across northern Australia but it has declined markedly since European settlement and is now abundant only on Barrow Island off the coast of Western Australia. Easily identified by its coarse grey fur and reddish 'spectacles' around its eyes, it is active only at night when it browses on the foliage of shrubs and the tips of spinifex; it does not drink even if water is available. By day it remains deep in the shelter of a spinifex clump or similar retreat.

Quokka

Quokkas are scarce on the mainland but abundant on Rottnest Island, which got its name from their day-time shelters.

The Quokka may look like any other wallaby but studies of its skull, teeth and DNA show that it belongs in its own separate group. Compared to most other wallabies, it has a rather short tail, very short rounded ears and long, coarse fur, grizzled grey with a hint of rufous. It is about 80 cm long and weighs about 4 kg. Confined to the far southwest of Australia, it inhabits dense wet heathlands and similar vegetation on the mainland but on the nearby off-shore island of Rottnest it also lives in arid scrub. Although ground-dwelling, it occasionally clambers into low shrubs to browse. Quokka populations have been much reduced on the mainland since European settlement but numbers have recently recovered and it is now locally common and on Rottnest Island it is abundant.

QUOKKA ISLAND
The Dutch mariner Willem de Vlamingh named Rottnest Island after the Quokkas. Landing in 1696, he described the quokka as 'a kind of rat as big as a common cat'. Rottenest is Dutch for rat's nest; it was later changed to Rottnest.

Social Organisation

The Quokka lives in family parties dominated by adult males. A creature of habit, its activities centre around a secure position in a thicket or similar sheltered location, to which it returns daily at dawn to sleep through the heat of the day. Such sites occasionally change in winter but are otherwise used year-round, and males sometimes fight for possession of the best ones.

Tree-kangaroos

Bennett's Tree-kangaroo (above) and Lumholtz's Tree-kangaroo (right) are readily distinguished by their different face-patterns.

Tree-kangaroos have evolved in the tropical rainforests of New Guinea and two species occur in far northern Queensland. Compared to terrestrial kangaroos, they have much stouter forelegs and shorter hindlegs. Tree-kangaroos are nocturnal. They feed mainly on leaves and fruits and spend much of their time high in the treetops but they frequently come to the ground and even cross open areas. They are usually very hard to spot but are occasionally betrayed by their long dangling tails. They sometimes occupy very small remnant patches of rainforest.

In the forest canopy, tree-kangaroos lack stealth and grace but not agility. Tree-trunks are descended tail-first, forepaws clutching and hindpaws scrabbling, and the last few metres to the ground are normally covered in a single twisting leap, landing on the hindfeet and facing away from the trunk.

Species Differences

Both Australian tree-kangaroos are solitary, strongly territorial in behaviour and about 1.5 m in total length. Lumholtz's Tree-kangaroo weighs 5–8 kg and is distinguished by a more or less plain tail and a conspicuous pale band across its forehead. Bennett's Tree-kangaroo is more thick set, weighing 8–14 kg, and is recognised by its plain greyish face and the conspicuous long whitish patch on the upper surface of its blackish tail.

SLOW BUT STEADY ON A LEAFY DIET

In response to the high-energy demands of a leaf-based diet, tree-kangaroos have a much lower metabolic rate than other kangaroos. They also breed more slowly, with the 45-day gestation period being longer than that of any other marsupial and joeys spending up to three years with their mother.

Wallaroos

Wallaroos may be seen in pairs or family groups, but seldom in large parties.

Wallaroos live in hilly, often rocky, country and to suit this terrain they are stockier and more muscular than most other wallabies and kangaroos. Their broad-shouldered stance, with elbows tucked in and wrists raised, is distinctive. Males are twice the size of females. The Black Wallaroo is confined to the escarpment country of Arnhem Land and the Antilopine Wallaroo to the wooded country of the Top End. Most widespread is the Common Wallaroo.

Two Forms: One Species

The Common Wallaroo (see map) occurs in two forms so different in appearance that they bear distinct common names — the Wallaroo in the east and the Euro in the west. Nearly as big as the grey kangaroos, the Wallaroo is usually dark grey-brown in colour and has a coarse shaggy pelt, whereas the Euro has shorter, more reddish fur. In both forms, the nose is bare, not furred. The Common Wallaroo usually occurs alone or in pairs, and seldom forms large mobs. It prefers a shady rock ledge or a shallow cave in which to rest and doze during the day. At dusk it moves downslope to graze in clearings, paddocks or adjacent grassy plains. It can survive indefinitely without water.

Why do Kangaroos Hop?

Some of the larger kangaroos can hop at 40 kph over short distances but speed is not the only measure of effective locomotion. It turns out that, over short distances, hopping is very efficient because a great deal of energy is conserved in the powerful springs formed by the muscles and ligaments of the kangaroo's massive hindquarters and this is recycled on ensuing hops.

The kangaroo's distinctive hopping gait is not necessarily faster than running but it is more efficient.

How do Roos Keep Cool?

Kangaroos and wallabies lack sweat glands and so they do not perspire as humans do in order to reduce their body temperature. Instead, they frequently lick their paws. As the moisture evaporates it absorbs heat from the body, resulting in a reduction of body temperature. By the same principle, you may shiver after swimming as water droplets take heat from your body .

Some kangaroos have the distinctive habit of licking their paws to keep cool.

Why Joeys Suckle from the Same Teat

Sometimes a new joey is born before its older sibling is entirely weaned, so both are suckling together for a time. In most mammals the milk a mother provides for her infant constantly changes in chemical composition to match the baby's changing needs as it grows. Remarkably a female kangaroo has the ability to deliver milk of significantly different chemical compositions through different teats simultaneously: one formula suited to a tiny infant, the other suited to her almost-weaned older joey.

A mother wallaby with her well-grown joey.

What's Special about Kangaroos?

Some kangaroos have evolved an unusual reproduction strategy known as embryonic diapause. Within a few days of giving birth, females are ready to mate again but the fertilised egg that results undergoes a few cell divisions before its growth is interrupted. So long as she has a joey in her pouch the new embryo is held in suspense but when the joey is finally weaned and quits the pouch it resumes normal development. This means that if one joey should die for any reason, she has another embryo in reserve. If conditions are good she can promptly switch her parental commitment from the older joey to her reserve embryo without first having to delay to find a mate. Kangaroos vary in their use of this tactic. For example, the Eastern Grey Kangaroo exhibits embryonic diapause but the Western Grey Kangaroo does not.

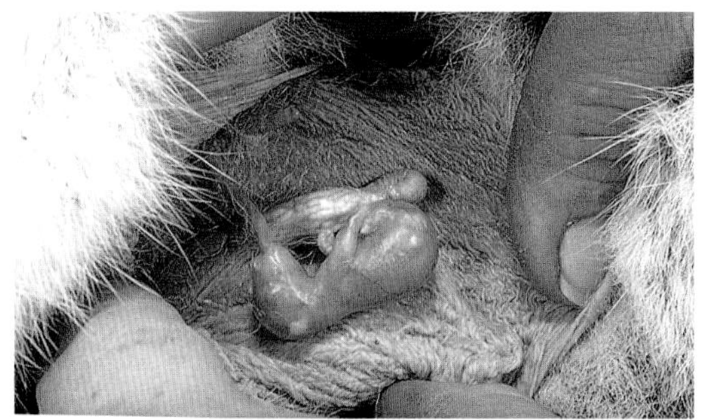

At first a newborn joey remains locked to its mother's teat.

NEWBORN ROOS

Even the largest kangaroo at birth weighs less than 2 g. Unaided it finds its way through its mother's fur to her pouch. Once safely inside, it clamps onto one of her four teats and remains there until ready to explore the world outside at about 9 months.

Red Kangaroo

Most male Red Kangaroos are easily identified by their reddish fur.

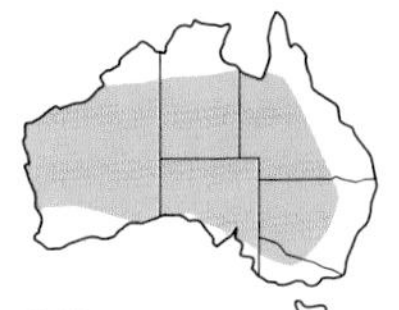

Over much of outback Australia, the large kangaroos are by far the most obvious native mammals, easily seen by even the most casual observer. This is especially true of the Red Kangaroo, which is the largest of all living marsupials: males sometimes reach 85 kg in weight and, rearing fully upright, can stand the height of a tall man. Females are often only half as big as males. Generally the Red Kangaroo is easily identified by its reddish fur and white underparts but females — and occasionally males, especially in the west — are bluish grey in colour. Widespread across the arid interior, wherever there is grass and at least a few trees to provide shade during the day, these big kangaroos generally favour open country, mainly where the average annual rainfall is 500 mm, leaving hills and rocky country to the wallaroos.

Social Habits

Red Kangaroos live in mobs that vary greatly in size depending on local and seasonal conditions, from family parties to herds several hundreds strong. They are partly nomadic, moving in response to local rainfall in search of green vegetation. Like all the larger kangaroos and wallabies, they feed mainly on grass and similar low vegetation, foraging mainly at night and resting up during the day. They come regularly to drink when water is available but they can also survive for long periods without it.

Grey Kangaroos

Western Grey Kangaroos (above) are extremely difficult to distinguish from Eastern Grey Kangaroos (right), but are usually slightly browner.

The grey kangaroos are probably the most widespread kangaroos in Australia. Nearly as big as the Red Kangaroo, the two species look very similar to one another but the fur of the Eastern Grey Kangaroo is distinctly grey in tone, while that of the Western Grey Kangaroo is much browner. On the whole, the Western Grey Kangaroo inhabits more arid country than its eastern relative. Both have hairy muzzles, a useful feature distinguishing them from most other kangaroos and wallabies.

Grey kangaroos usually occur in mobs of 50 or so. These consist mostly of females and their young and are visited by mature males during the breeding season. They eat mostly grass, foraging mainly during the night and resting up in the shade during the day.

UNUSUAL TEETH

To grind grass to a pulp, kangaroos have two large grinding teeth, or molars, at the back of each jaw and on each side. As the animal ages these slowly creep forward so that the point of heaviest wear steadily shifts from the front set to the back set, enabling the whole assembly to last longer.

Eastern Grey Kangaroo

The Eastern Grey, which is confined to the eastern part of Australia (see map), can be easily confused with the smaller Common Wallaroo and further west, especially on the plains of western New South Wales, with Western Greys, which may even occur in the same mobs. Western Greys occupy most of southern Australia.

Pademelons

Very common in the southeast, the Red-necked Pademelon is the best known of all pademelons. Right: a Tasmanian Pademelon.

Pademelons are kangaroos that live in rainforests or similar tall, dense woodlands. To be more precise, they inhabit the edges of such forests, for they generally forage in clearings or adjacent grasslands at night and spend the day deep in the forest, sometimes commuting several kilometres between daytime refuges and nocturnal feeding grounds. They are solitary animals: even though groups may gather when foraging, such congregations appear to form entirely by chance. If alarmed or disturbed, they scatter instead of staying bunched together like many kangaroos.

Species Differences

All three species are similar in general appearance and size — about 1 m long and 5 kg in weight. The Red-necked Pademelon is perhaps the most common in the heavily populated southeast. The Tasmanian Pademelon is now confined to Tasmania but it was once widespread in Victoria. The distribution of the Red-legged Pademelon widely overlaps that of the Red-necked Pademelon but also extends much further north to the tip of Cape York Peninsula. The Red-legged Pademelon is also more nearly confined to dense rainforest. Both the Red-necked and the Red-legged Pademelons have strong chestnut accents in their fur: in the former these are most obvious across the shoulders, while in the latter they are on the hips.

KEEPING TRACK

Pademelons are creatures of habit and often use the same routes daily as they commute between sleeping and feeding areas. Narrow tunnel-like runways in thickets and dense undergrowth sometimes betray their presence.

Swamp Wallaby

The Swamp Wallaby normally lives in dense shrubbery but emerges onto open grass to feed.

Although the Swamp Wallaby seems, on casual examination, to be fairly typical of many other wallabies and small kangaroos in general appearance, it is actually very distinct in several genetic, anatomical and behavioural characteristics. Males, for example, have 11 chromosomes and females 10, whereas all other kangaroos and wallabies have 16.

The Swamp Wallaby is usually identifiable by its dark colour alone but it also moves very differently from other wallabies, with its head held very low and the tail extended rigidly behind it as it hops. Its total length is about 1.5 m, approximately half of which is tail; females are slightly smaller than males. The dark tail sometimes has a pale tip.

Habitat, Behaviour and Breeding

Despite its name, the Swamp Wallaby is not restricted to swamps but it does favour dense thickets, moist gullies and similar heavy vegetation for shelter during the day; in Queensland it is especially common in brigalow scrub. Like other small forest wallabies it is mostly solitary in behaviour and forages on leaves and other vegetation as well as grass. There seems to be no particular breeding season. The joey leaves its mother's pouch at about nine months of age but remains to some extent dependent for another six or seven months.

Nailtail-wallabies

The tiny but distinctive horny tail-tip is clearly visible on this Bridled Nailtail-wallaby.

The three species of nailtail-wallabies are characterised by a horny, nail-like tip to the tail, function unknown. The Bridled Nailtail-wallaby once occurred along the western slopes of the Great Dividing Range from around Charters Towers to the Murray River but it is now confined to a few localities in Queensland. It gets its name from a narrow 'bridle' of white fur that extends from the base of each ear along the nape and around the back of each dark reddish shoulder. Its close relative, the Northern Nailtail-wallaby, occurs on Cape York Peninsula, the Top End and the Kimberley (see map). The third member, the Crescent Nailtail-wallaby of Western Australia, is extinct.

Habitat and Habits

Nailtails are, or were, inhabitants of grassy woodlands. When hopping, they seem to 'pedal' their forelimbs in a distinctive rotary motion that suggests their locally common name of 'organ-grinder'. Shy and mainly solitary, they spend the day in thickets of deep shade and emerge in the evening to feed on grass, roots and low foliage. Sometimes, especially in winter, they pause first to bask for a while in the setting sun. When disturbed or alarmed, they often take refuge in a hollow log or similar shelter.

Rock-wallabies

The Yellow-footed Rock-wallaby (above) inhabits much drier country than its close relative the Unadorned Rock-wallaby (right).

As their name suggests, all rock-wallabies are strongly associated with boulder-strewn hillsides, rocky outcrops and such places. The striking thing about the rock-wallabies is the large number of species and the restricted distribution of each of them. About 15 species are usually recognised. Seldom do their ranges abut and only in the Kimberley do several species occur together. They have a number of distinctive features suiting them to a life among rocks: for example, their broad hair-fringed hindfeet function like hob-nailed boots and their tail — often carried arched over the back — is far more slender and less tapering than other kangaroos, and is used to balance rather than as a prop. Very much at home among boulders, they scramble up cliffs and bound from rock to rock with extraordinary agility.

Yellow-footed Rock-wallaby

Clad in shades of grey, yellow and orange, with a banded tail, the Yellow-footed Rock-wallaby is among the most colourful and strongly marked of all the marsupials and at one time it was hunted for its beautiful pelt. It is now fully protected. Although vulnerable, it remains locally common, especially in the Flinders Ranges of South Australia and the Adavale region of southwestern Queensland. It lives in groups and feeds mainly on grass.

WOODLAND THUMPERS

Presumably as a signal of danger to others of their group, many kangaroos, wallabies and pademelons stamp the ground with their hindfeet when disturbed, producing a hollow thumping sound.

PLACENTAL MAMMALS

What is a Megabat?

Bats fall into two major groups. The 'megabats' or megachiropterans are generally rather large: some have wingspans of nearly 2 m. Megabats eat mostly fruit and rely on vision to find their way around. The second group, the 'microbats' or microchiropterans, are generally very small — typically about mouse-sized; they generally feed on insects captured in flight and use echolocation to navigate and find food.

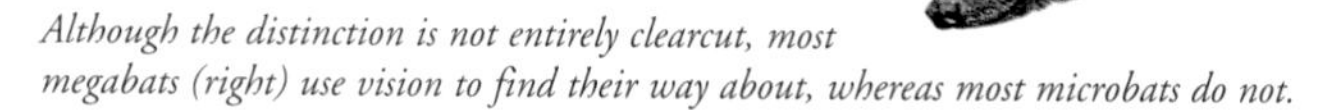

Although the distinction is not entirely clearcut, most megabats (right) use vision to find their way about, whereas most microbats do not.

Are Bats Blind?

The common expression 'blind as a bat' is not true. No bats are blind, although it is true that most rely very little on vision for information about the world around them. Instead, the bat's main claim to fame is its dazzling virtuosity in the arcane, hi-tech world of biosonar.

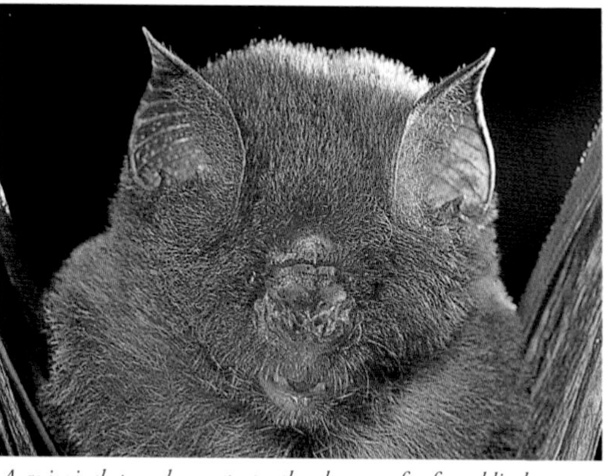

A quizzical stare demonstrates that bats are far from blind.

Microbat calls are very loud and very high-pitched.

SOUND ENGINEERS

A bat uttering, say, 10 sound pulses per second while searching for prey may raise the number of emissions to 200 per second as it closes in on its target. Humans cannot hear sounds above 20 kHz, while bats have been recorded uttering pulses at frequencies of 150 kHz or more.

What is Biosonar?

If you stand in a completely dark room and shout, you discover a great deal about the size and shape of the room merely by listening to the echo bouncing off the walls. You have just used biosonar to explore your surroundings. Bats do exactly the same thing to navigate and catch food but they are vastly better at it.

The biosonar abilities of bats are highly sophisticated. To begin with, bats are most interested in insects, which are small. To bounce a sound off a small target, the signal or pulse needs to be of a very high frequency. The smaller the target, the higher the frequency needed to detect it; and the higher the pitch, the sharper the 'image' formed by the resulting echo.

Megabats roost in the open in 'camps'.

Duration also matters: a very brief sound pulse pinpoints the target more accurately than a long one. Moving targets add further complications: a single 'image' tells the bat about location but nothing at all about course and speed. To plot an interception course it needs several 'images' in rapid succession, the quicker the better, especially if the target is flying erratically.

A bat's grotesque facial equipment, which may involve complicated structures in the ear or fantastically elaborate folds and wattles around the nostrils, can be viewed as devices for the delicate and very precise control of both the outgoing signal and the returning echo.

Bizzare ears and nose-leaves are actually hi-tech sound controllers.

DIVERSITY

The world's bats number about 1,000 species; in other words, about one out of every four of the world's mammal species is a bat of some kind.

Flying-foxes

Megabats, such as these Black Flying-foxes (above) and Grey-headed Flying-foxes (right), characteristically furl their wings about their bodies.

Flying-foxes are megabats, many with a wingspan of a metre or so. Generally they are confined to the forests and woodlands of the coastal strip. In Victoria and most of New South Wales only the Grey-headed Flying-fox (see map) and the widespread Little Red Flying-fox are common, although the latter tends to favour inland districts. The Black Flying-fox occurs on the east coast from about Lismore northwards and is the common species across the Top End and the Kimberley to about Shark Bay in the west, while the Spectacled Flying-fox is common in the northeastern tropics.

INFANT CARE

Most young bats are born in October, although breeding can take place at any time. The single infant clings to its mother for the first 3 or 4 weeks of life; thereafter it is left at the camp while its mother forages.

Grey-headed Flying-fox

Readily distinguished by its shaggy fur, reddish or yellowish collar, and greyish or whitish head, the Grey-headed Flying-fox can often be heard making harsh, squealing noises in the fig trees of city parks at night in coastal cities such as Sydney. At dusk or just afterwards these large bats leave their camps and stream across the sky, sometimes in hundreds of thousands; they may travel 50 km or more to reach a good food source. By day they retreat to the camps in secluded forests, often in gullies near water. Here they congregate in large numbers to sleep during the day.

Ghost Bat

The Ghost Bat is carnivorous, preying on mice, songbirds and skinks as well as other bats.

The Ghost Bat is the only Australian member of a small family that is otherwise widespread in Africa and Asia. It preys on other bats and small creatures such as mice, small birds, frogs, lizards and large insects. Although it has echolocation abilities, it does not always use them: owl-like, it often waits patiently on a low perch until something flies or scuttles by, then swoops to intercept. It often carries its prey to a favoured feeding perch in a tree or a small cave, the presence of which may be sometimes detected by the litter of discarded animal parts that accumulate below.

The Ghost Bat is a large microbat, about 13 cm in total length, with pale fur and large ears and eyes. Although it hunts alone, it gathers in groups of up to several hundred to roost in caves, mine shafts and similar places during the day.

Distribution

Around the time of European settlement, the Ghost Bat appears to have occurred across most of the northern half of Australia, including the interior, and extending southward perhaps as far as the Flinders Ranges. However, there seems to have been a steady decline ever since; it remains common only in the Northern Territory and its distribution is decidedly patchy elsewhere.

FLYING MAMMALS

Although several mammals use membranes to glide, bats are the only mammals capable of flapping these membranes and sustaining flight.

How do Microbats Differ from One Another?

Microbats are extremely difficult to tell apart, even in the case of captured specimens held in the hand. In learning to recognise the various families, it is firstly worth noting that the membrane that extends between a bat's hindlegs is technically known as the uropatagium. You can make a useful start to identification by noting the condition of the uropatagium, the ears and the presence or absence of a nose-leaf.

Ear and nose-leaf structures are crucial in microbat identification.

Of Australian microbats, only the Ghost Bat, family Megadermatidae, is reasonably easy to identify because of its large size, large eyes, large ears and pale fur. The horseshoe bats in the family Rhinolophidae and the leaf-nosed bats, family Hipposideridae, both have very large ears and elaborate nose-leaves. The sheathtail bats in the family Emballonuridae have a moderately long tail that seems to penetrate the uropatagium from below, and the freetail bats, family Molossidae, have a 'free' tail that extends considerably beyond their uropatagium.

All other Australian microbats belong to the fifth family, Vespertilinonidae. This extremely varied family lacks any obvious distinguishing external features, although in Australia any small bat that lacks a distinct noseleaf and has a tail that extends only, more or less, to the edge of its uropatagium is almost certainly a vespertilinionid of some kind.

A Southern Freetail Bat in flight reveals its distinctive tail.

SPERM STORAGE

In virtually all mammals, sperm is viable for only a few days at most but many bats have the remarkable ability to store it for months at a time, usually throughout the winter months for use in the spring.

How do Bats Benefits Plants?

Among the smallest of the megabats are the mouse-sized blossom-bats. Two species are common in Australia: the Eastern Blossom-bat in Queensland and the Northern Blossom-bat in the Top End (although they overlap considerably in tropical Queensland). Roosting in forest and foraging mainly on heaths, these small bats have long, brush-tipped tongues. A diet of almost exclusively nectar and pollen makes them important pollinators of several plants.

Blossom-bats are tiny and feed mostly on nectar.

Where to see Bats

One of the best ways of observing bats is to visit a roosting or maternity cave but be aware that bats are most vulnerable at their roosts, especially if they are hibernating or in torpor during winter. In a few moments of agitation and flying around bats awoken by visiting humans may use up as much energy as would keep them alive for weeks if they remained undisturbed. Unable to feed until spring, you may inadvertently run them out of fuel and they may then starve, so always visit such caves only with great care and preferably only with an experienced and knowledgeable guide.

Bats of many species typically roost or rear their young in caves.

An Eastern Tube-nosed Bat at rest.

Eastern Tube-nosed Bat

The Eastern Tube-nosed Bat is a close relative of the flying-foxes but it is very much smaller — not much more than 12 cm in total length — despite being a megabat. It is common along the Queensland coast from around Coollangatta to Cape York. It is easily identified by numerous, and highly variable yellowish spots on its wings and similar markings on its ears; these make it look very like a dead leaf at its daytime roost. It eats fruit and can sometimes be a nuisance in orchards. Relatively sedentary and more or less solitary in behaviour, it sometimes roosts in the same fruit tree from which it has foraged the night before. Its distinctive bleating call may be uttered only by males.

UPSIDE DOWN

Most bats at their roosts dangle upsidedown from their toes, which are strongly clawed. While megabats furl their wings closely around their bodies like a cloak, Dracula-style, the smaller bats merely fold their wings along their sides.

Large ears help identify this bat species.

Eastern Horseshoe-bat

The Eastern Horseshoe-bat hunts within dense forests, from the canopy to the undergrowth within centimetres of the ground. Its normal flight is slow, erratic and fluttery. Flying moths and beetles are favoured prey. These are often taken to its daytime roost for dismemberment at leisure. Most communal roosts contain only about 10 to 50 individuals, although a thousand or two have been recorded in some. Road culverts or disused buildings, which are merely shaded rather than pitch-dark, are commonly utilised.

The Eastern Horseshoe-bat is common along the entire east coast of Australia from Victoria to Cape York. In the southern States it is recognisable as the only bat with extremely large ears and an obvious, very elaborate nose-leaf. In the tropical north, however, half a dozen species with very large ears, elaborate nose-leaves and similar lifestyles and behaviour belong to a closely related group, the leaf-nosed bats.

The forest-dwelling Eastern False Pipistrelle.

Eastern False Pipistrelle

The Eastern False Pipistrelle inhabits eucalypt forests. Its distribution includes Tasmania and the southeastern mainland of Australia about as far north as Brisbane. A close relative, the Western False Pipistrelle occurs in the old-growth forests of far southwestern Australia. The Eastern False Pipistrelle may be much more common than available information indicates but little is known of its life history. Its rather long wings and fast, direct flight suit it to patrolling the lower levels of the forest rather than the more cluttered environment of the canopy. It roosts in hollow trees and the sexes are usually segregated, at least for part of the year. It spends much of the winter hibernating and males of the species are among those that produce sperm in autumn but store it through the winter for use in early spring.

The "freetail" is a distinctive feature of this species.

Southern Freetail Bat

As the name suggests, the freetail bats are characterised by the fact that their tail extends freely for much of its length beyond the membrane between the hindlegs. They resemble sheathtail bats in having relatively long, slender, tapered wings adapted for high-speed hunting in open, uncluttered environments but most are even more nimble on the ground where they often chase ground-dwelling invertebrates such as spiders and cockroaches.

The Southern Freetail Bat's body is noticeably flattened in profile, with a broad, shallow skull. These features enable it to squeeze into narrow crevices and quite small cavities in trees to roost by day. Such roosts seldom contain more than 10 individuals.

AUSTRALIAN BATS

Bats are relative newcomers to Australia — the earliest immigrants did not arrive until perhaps 20 million years ago — and they are not especially diverse by global standards. Even so, there are at least 60 Australian species, and more are still being discovered.

Lesser Long-eared Bats often fall prey to cats.

Lesser Long-eared Bat

Long-eared bats have very large ears, large eyes and a small, simple nose-leaf. There are several species, all very difficult to tell apart, but the Lesser Long-eared Bat is one of the most numerous, by far the most widespread and often common in towns and cities where it may roost in the roofs and attics of houses. The wings are short, broad and well adapted to slow, fluttering, highly manoeuvrable flight in confined spaces. It often forages in shrubbery close to the ground and much of its prey is caught by gleaning it from leaves or by simply perching and waiting for it. This habit makes it especially vulnerable to capture by cats.

The Lessert Long-eared Bat may roost either alone behind slabs of loose bark on trees, in natural tree cavities or in buildings, or it may gather in groups of a hundred or more, often in the company of other bat species.

MATERNITY ROOSTS

Pregnant female long-eared bats tend to congregate in clusters of a dozen or so in special 'maternity' roosts to rear their young. Twins are common and, while very young, both are sometimes carried by their mother as she hunts. They are weaned at about 6 weeks, by which time they can fly and make their first tentative hunting trips alone.

The diminutive Little Forest Bat.

Little Forest Bat

This tiny bat is a little less than 7 cm in total length and weighs only about 3 g. Although rarely recognised, it is abundant in the forests and much of the populated areas of southeastern Australia. It feeds on small moths and other tiny insects, usually captured below the forest canopy and swallowed on the wing, although sometimes larger prey is taken to a perch to be eaten at leisure. Preferring to roost in tree hollows, it is also often found in the eaves or attics of houses. Roosts are segregated by sex: males usually rest alone, females in small groups of 10 to 20.

The Little Forest Bat is one of nine very similar species, all of which are exclusively Australian: often several species are found together in the one habitat. Most mate in autumn but young are born the following summer. Little Forest Bats spend most of the winter asleep in their roosts and may not emerge to feed for weeks at a time.

This species exhibits uniquely folding wings.

Common Bentwing-bat

Bentwing-bats are all but impossible for the casual observer to identify to species but the group is distinctive in that the last joint of the third (middle) finger is several times the length of the preceding joint, producing an extended wingtip that is folded backwards when at rest. These bats have short muzzles and high-domed skulls. The Common Bentwing-bat is abundant in eastern and northern Australia, and it also occurs widely in Africa, Europe and Asia. It favours open environments for hunting, capturing its prey in sudden swoops from fast, level flight. Roosts are in caves, mine-shafts or buildings. Bats may be scattered or in large and tightly packed huddles. In springtime, females usually select the warmest and most humid caves available in the district as special 'nursery' roosts. Around December mothers normally produce a single baby, which gathers with others on the roof of the cave when its parent is foraging.

The slow-flying Gould's Wattled Bat.

Gould's Wattled Bat

There are several species of wattled bats in Australia, all extremely difficult to tell apart but Gould's Wattled Bat is one of the most numerous and widespread of all Australian bats, with a distribution that extends across the continent from Tasmania to the Top End. It is about 7 cm long, a velvety chocolate brown to black above and dull brown below. It lives in eucalypt forests and flies slow and low, rarely venturing above 20 m in the air. By day it roosts alone or in small groups in the hollows of large gum trees, and occasionally in caves or the attics or basements of houses. Like several other kinds of bats, females change their roosts in spring and congregate in special 'nursery' caves to give birth and rear their young. Most births are single but twins are also common and when very young, these are carried by the mother in flight.

A TINY CONTENDER

A contender for the world's smallest mammal is a bat. Only described in 1974, Kitti's Hog-nosed Bat from Thailand may even be the smallest warm-blooded animal — only a few hummingbird species rival it for minuteness. Individuals 3 cm long and with a wingspan of 15 cm can weigh 1.5 g: that is a bit less than the weight of a 5c coin.

What is a Rodent?

The Western Mouse is a typical native rodent.

The Pale Field Rat is a relative newcomer.

Rodents, the group to which the familiar rats and mice belong, are mostly very small in size but almost half of all the world's mammals — and about one quarter of Australia's land mammals — are rodents of some kind, and they occupy all continents except Antarctica.

One of their most distinctive features is the presence of two long, backward-curving teeth in the centre of the upper and lower jaws. The inner surface is mostly dentine but the front surface is clad in extremely hard enamel. These four teeth are never shed and grow continuously throughout the animal's life. They meet very precisely in such a way that the lower teeth grind against the inner surface of the upper teeth, wearing away the comparatively soft inner dentine more quickly than the hard outer enamel. In so doing, the assembly works much like a knife sharpener, always keeping the biting edge of the teeth honed to a razor sharpness.

Although there are, of course, exceptions, fecundity is another prominent characteristic of rodents. The onset of sexual maturity is often early — mere weeks in many species. Large litters of a dozen or more are common and there may be three or four in a season. The average annual output of young per female is therefore much higher than in most other mammals.

The Fawn-footed Melomys is a member of the mosaic-tailed rat group. Its ancestors arrived 15 million years ago.

Who were the First Rodents?

The first wave of rodents to invade the isolated continent of Australia are known as the 'old endemics'. Although they had been scurrying about in the world for around 60 million years, they did not reach our shores, presumably from Indonesia, until less than 15 million years ago. Even so, they prospered to such an extent that these so-called 'old endemics' now have diversified so greatly that there is room for only a sample in this book. They occur in every corner of Australia, and include species that live in trees, on the ground and even in the water. They also occupy all types of habitats, including deserts and rainforests.

Despite its name, the Brush-tailed Tree-rat is almost equally at home in the trees or on the ground.

The 'old endemics' can be identified from females, which have only four or six teats, while all others have eight to 12. Prominent 'early invaders' are the tree-rats (*Conilurus*), the stick-nest rats (*Leporillus*), the hopping-mice (*Notomys*), the pebble-mound mice (*Pseudomys*), the rock-rats (*Zyzomys*) and the mosaic-tailed rats (*Melomys*). Most of these are found nowhere else in the world.

Australia received a second influx of rodent immigrants — the so-called 'new endemics' — sometime within the last few million years. Although these have evolved into distinct species, they are closely related to the familiar introduced pests and are included in the same genus, *Rattus*.

The Sandy Inland Mouse is widespread in the arid interior.

WHO WERE THE LATECOMERS?

A third influx occurred when Europeans arrived, bringing with them the House Mouse, the Brown Rat and the Black Rat as accidental baggage. Although the two introduced rats have not penetrated far inland (except in the Riverina), the House Mouse now lives in every corner of the continent.

This rodent is common in the snow country.

Broad-toothed Rat

Just under 30 cm in total length, the Broad-toothed Rat is most common at high altitudes, especially the subalpine heaths of the Snowy Mountains and the Australian Alps, but it also occurs close to sea level in the Dandenongs and Otway Ranges and in Tasmania. Wherever it occurs, the climatic regime is cool in summer, cold in winter, and with a high rainfall. It favours areas of dense undergrowth, where winter snow lies not too snug on the ground, leaving runways through the vegetation beneath, which enable the animal to continue its activities through the winter. At such times several individuals often share a single nest but in summer each tends to build its own. The usual litter consists of one or two young, and females may bear two litters per year.

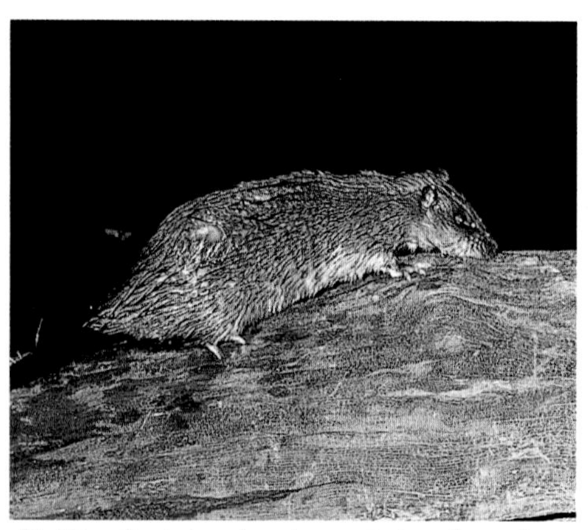

The Water Rat is the only truly aquatic Australian rodent.

Water Rat

As the name suggests this rat is usually found near water — fresh, salty or brackish — and much of its food is taken below the surface. It is widespread in coastal regions of Australia, as well as much of the eastern interior; it also occurs in New Guinea. An excellent swimmer, it eats some plant material but is mainly a predator on small aquatic animals of all kinds. Much of its foraging is done at night but it is often active before the sun goes down. About 50–60 cm in total length, its thick waterproof coat is usually grey above and dull orange below, and its thick, hairy tail usually has a white tip. Strongly territorial, the Water Rat is mostly solitary except for when mothers are accompanying their growing young. Females often bear up to five litters of one or two young per year.

The Giant White-tailed Rat has an awesome bite.

THE LITERATE RAT

The Giant White-tailed Rat is a frequent raider of homes in northern Queensland, where it seems to show particular fondness for canned condensed milk. Weighing about 1 kg, the rat has no trouble tearing into cans with its powerful jaws, but the puzzle is how it knows which cans contain milk (locals claim it reads the labels!).

Giant White-tailed Rat

The Giant White-tailed Rat is mainly a rainforest inhabitant but it also occurs in melaleuca swamps, dense eucalypt woodlands and similar forested country. It spends much of its time in trees but it also fossicks on the ground. Its diet is varied: it tears dead branches apart searching for wood-boring insects and their larvae, as well as eating fruits and seeds, its strong teeth and jaws enabling it to gnaw open many fruit seeds too large and hard for most other forest rodents. Usually active by night, it sleeps during the day in the cavity of a tree or sometimes in a burrow under a fallen log or some similar site. Most breeding takes place in the rainy season when females give birth to two or three, occasionally four, young. These are left in the burrow when she forages. Within about three months they reach independence.

This tree-rat is confined to highland rainforest.

Golden-backed Tree-rat

The distinguishing features of the Golden-backed Tree-rat include a long, hairy, white-tipped tail and a distinctive band of reddish or dull orange fur running down the middle of the back. Now effectively confined to Arnhem Land and the Kimberley in country too rugged for livestock, it once had a more extensive distribution across the tropical north, occupying habitats as diverse as blacksoil plains, palm groves, vine thickets, rocky escarpments and even coastal beaches. It is largely nocturnal and lives in pairs or small groups. It seems equally at home on the ground as in trees and relies heavily on fruits, flowers and termites for food. In the wild, breeding seems most prevalent in early spring but captive females can rear young at any time of year. Females have four teats but twins make up the most common litter. The offspring are weaned within seven weeks.

Bush Rats have a distinctive bounding gait.

Bush Rat

Of the same genus as the introduced rats and of similar size and general appearance are numerous native species, including the Bush Rat. Although one of the commonest and most widespread of all native rats in coastal areas, especially in the heavily populated southeast, the Bush Rat is difficult to see because it is mainly nocturnal and usually remains in dense cover. On the whole, it prefers bushland to grassland, where its diet includes a wide range of fungi, seeds and other plant materials as well as — unusually for a rodent — a very substantial number of insects.

For much of the year, the Bush Rat is mainly solitary, both sexes independently occupying territories or home ranges that nevertheless often overlap extensively. Breeding may occur at any time but peaks in spring, especially in the south. Both sexes can breed at about four months, the usual litter size is five, and three or even four litters may be raised in a favourable season. However, few Bush Rats survive their second winter.

The attractive Mitchell's Hopping-mouse.

Mitchell's Hopping-mouse

Mitchell's Hopping-mouse averages about 26 cm long, of which about 15 cm is tail. It is nocturnal and by day sleeps in a burrow in the ground. It feeds largely on seeds but much of its water requirements are obtained from green shoots and the bodies of insects with which it supplements its diet. Little is known of its breeding behaviour but the usual litter of three or four young are weaned at about five weeks.

There are about 10 species of hopping-mice distributed across the arid parts of Australia, especially across the southern interior, where several species appear to occur according to habitat: Mitchell's Hopping-mouse, for example, inhabits mallee woodland, whereas its close relatives, the Fawn Hopping-mouse inhabits gibber plains; the Dusky Hopping-mouse, sand dune country; and the Spinifex Hopping-mouse, spinifex sand-plains.

Western Pebble-mound Mouse

Western Pebble-mound Mice are clever engineers.

The Western Pebble-mound Mouse is of great interest because it builds permanent structures in the form of mounds of pebbles, which can be quite obvious and extend over several square metres. Active colonies are characterised by volcanic rims of small stones around each entrance to the underground burrow system. Pebbles may weigh up to 10 g — very nearly the weight of the animals themselves. They lift these pebbles in their jaws and wrestle them into position with their forepaws, yet the point of all this labour remains uncertain.

The generic name for pebble-mound mice is *Pseudomys,* which means 'false-mouse' and all 23 species of these native mice look very like the introduced House Mouse. Although almost every part of Australia is inhabited by at least one 'pseudomouse', there are seldom more than two at any particular locality. Only a handful of these demonstrate anything like the mound-building skills of the Western Pebble-mound Mouse.

Common Rock-rat

The Common Rock-rat's tail is very fragile.

The Common Rock-rat is widespread in rocky country across tropical northern Australia. It is about 25 cm long, of which a little over half consists of tail. The tail is unusual in that the skin and fur is very fragile, easily lost or damaged and withers away rather than regrows so that many individuals have stumpy or shortened tails. Where it occurs it is often abundant, although difficult to observe because of its nocturnal habits: small heaps of chewed seeds on rock ledges often betray its presence.

This species tolerates a much wider variety of habitats than its other close relatives and is therefore much more widespread. It has local populations across the tropical north. Rock-rats are generally rigidly confined to rocky habitat and each of the regions of major rocky ranges — the Kimberley, Arnhem Land, Carpentaria and the Centre — has its own local species.

Introduced Mammals

Fallow Deer

Fallow Deer need access to water.

A number of species of deer have been introduced to Australia at one time or another but none with any great success, although six species currently have self-sustaining populations. Most of these are extremely local but two, the Sambar and the Fallow Deer, are somewhat more widespread, especially in Victoria and Tasmania.

About the size of a domestic goat, Fallow Deer are usually encountered in small groups or herds. Originally imported from England around 1850, they did well in Tasmania, where the total population is now about 8,000, but rather less well in most of the numerous places on the mainland where they were also released around the same time. Fallow Deer are primarily grazers, although they tend to spend more time browsing on trees and shrubs in winter. Favouring open, grassy country in which to forage, they also require nearby forest and woodland for shelter.

Dingo

Pure-bred dingoes are becoming rare.

A typical Dingo is sandy fawn in colour with white points, upstanding ears and a bushy tail but interbreeding with feral domestic dogs is widespread and the resulting crosses may show a wide range of features. Interbreeding is so common that it seems the Dingo's days are numbered, the breed extinguished by dilution. The original Dingo is a primitive dog that arose in India and was spread by East Asian traders and travellers throughout the region, apparently arriving in Australia some 4,000–6,000 years ago. As predators, Dingoes are unusually flexible. Although they usually live together in packs, when it comes to foraging they are equally adept at splitting up to hunt small prey alone or joining forces to hunt larger game in concert. Which is the better strategy depends on a wide range of very variable local factors, so the Dingo's ability to switch readily from one to the other means that it can survive in many different habitats.

The versatile Red Fox occurs in many habitats.

Red Fox

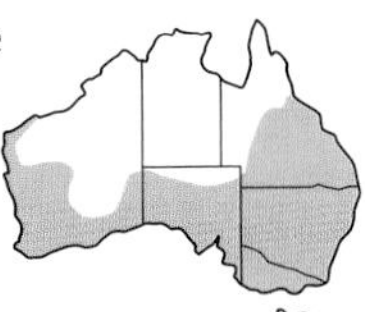

From its natural home in Eurasia, the Red Fox was deliberately introduced during the late nineteenth century in order to emulate the English sport of fox-hunting in Victoria. It spread rapidly and now occurs almost everywhere in southern Australia except the harshest deserts. It does not occur in Tasmania nor the northern tropics. The Red Fox is an efficient predator, preying mainly on rabbits and other small mammals, but it is also a skilful scavenger, its diet extending from insects and berries to carrion. Like the Dingo, its versatility in both diet and hunting technique allows it to prosper in a wide range of habitats although, unlike the Feral Cat, it is disadvantaged in deserts because it needs access to drinking water. It usually forages by night, but it is often active by day as well, especially when food is scarce.

Feral Cats adapt easily to arid environments.

Feral Cat

As a supremely efficient and adaptable predator, there seems little doubt that Feral Cats have played a key role in the severe damage done to the Australian native fauna, especially small mammals, since European settlement. The exact extent of that role, however, remains uncertain. Although in times of plenty Feral Cats prey heavily on rabbits, their diet also includes — especially during drought — a wide range of small marsupials as well as birds, reptiles and large insects. Because cats do not need to drink, they can survive in the desert as readily as in a forest and their distribution is effectively throughout Australia. Eradication or control is extremely difficult, partly because their essentially solitary nature means they present an extremely diffuse 'target' for almost any conceivable control measure that might be devised. Also, control measures that do not threaten domestic pets have so far proved elusive, although a variant of feline herpes virus, against which pets can be vaccinated, is being studied.

A RECENT INTRODUCTION?

According to Aboriginal lore, cats have either always been present in the central deserts or were believed to have come from the west.

Rabbits cause severe soil erosion.

European Rabbit

The original home of the European Rabbit was Spain and Morocco but its domestication and its widespread introduction elsewhere in the world began with the Romans at least 2,000 years ago. Perhaps the worst single cause of damage to the Australian environment, the European Rabbit was brought to Australia in 1858 and over the next half century or so spread vigorously until now it is distributed virtually throughout the continent south of the Tropic of Capricorn. In some regions it is still spreading northward.

The European Rabbit feeds on grass and lives in burrows that it digs in the ground. Usually, it lives in communities, so that its burrows tend to be densely clumped to form groups called warrens. Close cropping of grass by dense communities of European Rabbits removes the topsoil's vital protection of vegetation, while their burrowing undermines it from below: the cumulative result of these activities is severe damage to Australia's fragile soils.

The swift Brown Hare.

Brown Hare

The Brown Hare looks very like a rabbit but it is larger and has much longer ears and legs; the ears are tipped with black. It is also very much more solitary in its habits. Notable for its speed, it can reach 50 kph for short distances. Hares eat grasses, clovers and other low vegetation, grazing at night and spending the day concealed and motionless in a thicket or among rocks; unlike rabbits, they do not burrow. Also unlike rabbits, which are born blind, naked and helpless, baby hares are born with a full coat of fur, their eyes open and able to move about almost immediately.

The Brown Hare was first introduced into Tasmania in 1830 and near Geelong sometime after 1859, mainly for the sport of coursing, in which hares are chased by greyhounds. It spread rapidly and is now widespread throughout eastern and southeastern Australia, especially in pastoral districts on the western slopes and tablelands of New South Wales and Victoria.

YOUNG HARES

Newborn hares are called leverets. About 3 days after birth, the mother takes each baby to a separate shallow scrape in the ground known as a 'form': come dusk they all meet at a central rendezvous to be nursed.

In Australia, feral horses are called Brumbies.

Brumby

Horses have been escaping from domestic stock since the first days of European settlement, and now herds of feral horses, or Brumbies, are common in the outback almost all across Australia. However, horses inhabit open grasslands and need frequent access to water, so they are rare or absent in desert areas and uncommon in woodlands and forests. Horses are social animals with a distinctive 'group-within-group' social structure. Much of their lives revolve around a family group consisting of a male, his harem of several females and their young. These groups themselves often band together to form herds that sometimes contain a thousand animals or more — although the numbers are usually much less than this.

Best known are the mountain Brumbies in the high country surrounding the Snowy Mountains and Australian Alps of the southeast. Here the breed apparently owes much of its strength and sure-footedness to a genetic contribution from large draught horses that were used in the early days of timber-getting.

Feral Goats are well adapted to arid regions.

Feral Goat

Goats were brought to Australia with the First Fleet in 1788 and feral populations were quickly established from escaped or abandoned livestock. Feral Goats are now common in the outback, especially in the arid and semi-arid parts of the eastern and western interior. Because they need access to water, they are sparse in desert areas and their numbers are controlled to some extent by Dingoes. Goats are versatile foragers and, at high densities, they inflict severe damage on the environment, causing erosion by breaking the topsoil with their hooves and consuming virtually all vegetation within about 1.8 m of the ground. Female goats are ready to breed at six months of age and can rear young every eight months thereafter, frequently giving birth to twins or triplets, especially as they get older. Breeding reaches a peak in spring but can occur at any time of year.

Feral Pigs inflict substantial damage on wetlands.

Feral Pig

The pig may have been brought to northern Australia by Indonesian traders and voyagers even before its arrival with the First Fleet in 1788 but, regardless of origins, for at least a century it has been numerous and widespread over much of the continent, especially in the east and north. Big, black and shaggy, Feral Pigs can be formidable and dangerous animals, even to humans.

Although they eat a great deal of plant material, especially underground roots and tubers, Feral Pigs are efficient predators, and they also eat any carrion they find. Needing access to water and shelter, they do not do well in arid regions and are absent or rare in much of South Australia, central Australia and Western Australia. Although mature males are solitary, pigs are generally social. Group sizes vary widely but herds of 50 or more are not uncommon, especially in the tropical north.

It is a myth that camels store water in their humps.

One-humped Camel

Camels were introduced into Australia as beasts of burden over a period extending from about 1840 to 1910 and for many years they were used in exploration, construction and transport. Their use declined as motor vehicles proliferated and many were simply released to become feral. The camels prospered under Australian outback conditions and — totalling perhaps about 100,000 individuals — camels are now widespread across the arid interior of Australia.

Adult male camels stand a little over 2 m at the shoulder and weigh about 1 tonne. Although they drink copiously when water is available, they can do without it for months. Camels normally live in close-knit groups of up to 40 or so, which occasionally combine to form much larger herds, usually under adverse conditions such as drought. Unlike most other feral species, they appear to have very little adverse environmental impact in Australia.

WATER IN THE HUMP

The popular belief that camels store water in their hump is a myth. However they can survive a 40 per cent loss in body fluid.

A Swamp Buffalo seeks shade under trees.

Swamp Buffalo

The Swamp Buffalo stands about 1.8 m at the shoulder and weighs about 1.2 tonnes. Its ancestoral form still survives in small numbers in parts of India and Nepal but the domestic variety is widely used in much of Southeast Asia as a beast of burden and in rice paddies.

There are two distinct forms of domesticated buffaloes: the so-called 'river' and 'swamp' buffaloes. Swamp Buffalo have wide-spanning horns that curve inward and backward, and are the strain introduced into Australia's Top End from Indonesia some time after 1825. Escaped or abandoned buffalo quickly spread across the floodplains of the tropical north. Until rigorous control measures were initiated in the mid-1980s, they caused severe damage to sensitive wetlands. A close relative, the smaller Banteng, is well established but very local on the Coburg Peninsula.

Feral Donkeys are numerous in the outback.

Feral Donkey

Like several other long-domesticated animals, the ancestor of the Feral Donkey, the African Ass, is endangered where it still exists in parts of Sudan, Somalia and Ethiopia. Its precise original distribution is unknown but it was domesticated in Egypt at least 6,000 years ago. Donkeys were brought to Australia in 1866 for use as pack animals but when they became superseded by motorised transport many were released or escaped and became feral. Australia's current total population of at least two million, scattered across the arid interior, may be the world's largest.

Donkeys resemble small horses but are distinguished by their much longer ears, upright mane and slender, tufted tail. Their social organisation is also approximately similar, being made up of family groups containing a dominant male, several females and a variable number of offspring. Such groups often aggregate to form larger herds.

What is Killing our Native Wildlife?

The small desert-dwelling Mulgara is at risk of extinction.

The Australian native mammal fauna has been severely damaged since European settlement. At least 17 species are now entirely extinct, another nine are now effectively confined to small offshore islands and the continued survival of many others is precarious at best. A distinct pattern emerges from these extinctions: whatever the causes, flying species, such as bats, have so far been immune but ground-dwelling species are disproportionately affected. Desert-dwellers have suffered more than rainforest dwellers and species weighing less than 5 kg are far more severely affected than those weighing in excess of 5 kg. In short, Australian mammal species most at risk are usually desert ground-dwellers weighing under 5 kg.

Causes are various and usually multiple, but — directly or indirectly — introduced mammals are prominent in the equation in most cases, and these reduce to two major factors.

The fox is a versatile predator of native mammals.

Direct Predation

Introduced carnivores, in particular the Dingo, Red Fox and Feral Cat, prey directly on small native mammals, as well as other vertebrates such as birds and reptiles. The question of whether foxes or cats have had the severest impact is an area of much current research. The cat, for example, seems strongly implicated in a spate of small mammal extinctions around or just after European settlement when there were no other predators and no significant land-use changes had yet occurred. Nevertheless, the Dingo and Red Fox have also played some part, and several extinctions can be attributed to predation from these introduced carnivores.

Indirect Factors

Several other introduced mammals exert their influence indirectly: by causing widespread changes to the environment or by competing directly with native mammals for food. These effects may be, in the long run, even more devastating than direct predation because adverse changes affect entire communities of animals rather than just a few particular species.

At high densities, goats can virtually denude a landscape.

At high densities in times of drought, both rabbits and Feral Goats are capable of removing virtually all ground vegetation from sensitive arid lands. Loss of this protective veneer of plants exposes the topsoil to erosion and to being blown away in dust-storms. Damage may be further compounded by the burrowing activities of rabbits or the sharp hooves of goats. Both pigs and buffaloes cause damage to sensitive wetlands, especially in the tropics, by trampling and their habit of wallowing in mudholes to protect themselves from sun and biting insects.

AWAY FROM HOME

Several of the introduced animals now feral in Australia are extinct as wild animals in their original homes, or nearly so. For example, Australia now has the only truly wild population of camels in the world.

Feral Pigs inflict severe damage on sensitive wetlands by trampling and wallowing.

Marine Mammals

Where do Marine Mammals Live?

All three groups of marine mammals are represented in Australian waters: the dugongs, the pinnipeds (seals and sea-lions) and the cetaceans (whales, porpoises and dolphins). The Dugong is a vegetarian inhabitant of shallow tropical coastal waters. Whales and their smaller relatives, the dolphins and porpoises, spend their entire lives at sea and never come ashore. Seals and sea-lions give birth and rear their pups on land, or ice, and frequently loaf about and sunbathe on beaches and rocky shores but they feed exclusively in the sea.

Although sea-lions may rest on land, they feed in the sea.

Who are the 'Fin-footed' Mammals?

These are the world's seals, sea-lions and walruses. They are often known collectively as pinnipeds. The word pinniped means simply 'fin-foot'. Unlike whales and dolphins, pinnipeds have four obvious limbs but these are strongly modified to form fins. In Australia, pinnipeds are represented by two very different families: the 'earless' seals and the 'eared' sea-lions and fur-seals.

SEA-LIONS AND SEALS

While sea-lions have distinct external ears and seals do not, the most obvious difference between the two groups is that sea-lions can use their limbs very nimbly on land, whereas seals can do little more than loll on the beach.

Do Whales Sing?

Whales and dolphins are noisy animals. Many dolphins locate prey by using a form of sonar — like bats do in air — and can utter sounds of such intensity that they can stun, and maybe kill, their victim at close range. Whales use sound to communicate. It may be a way of finding a mate or keeping in touch with one another. The Humpback Whale in particular makes the longest and most complicated sounds of all animals. Ranging from deep mournful moans to shrill squeals, some of their 'songs' can be detected by other humpbacks over tens or even hundreds of kilometres.

Why do Whales Beach Themselves?

Nobody really knows but one plausible hypothesis involves navigation by using the Earth's geomagnetic field. This field is influenced by surface features, such as coastal cliffs and nearby mountain ranges, resulting in contour lines rather like those on a topographical map. Whales navigate by following such contours. Sometimes local geology is such that the lines of magnetic force happen to strike a particular coast at a right angle. At such places, it might be that whales, following a particular contour line, simply follow the 'path' right up onto the beach. It is difficult to test such a notion but some degree of correlation between known geomagnetic anomalies and strandings has been found and the idea might help explain why whales sometimes promptly beach themselves again after having been refloated.

A pod of beached whales.

Whale-watching

Whale-watching is a growth industry and there are a number of places along Australia's coast, including Airlie Beach and Hervey Bay, where tourist operators offer trips out to sea in small boats specifically to watch whales. The best times are July to September. Legislation is in place to minimise harassment. In general, all craft must stay more than 100 m from any whale sighted; if there are more than three boats, then all newer arrivals must remain outside a 300 m radius.

Whale-watching from boats has become a popular tourist activity.

TAIL MOTION

Whales and dolphins progress through the water by up and down strokes of their powerful horizontal tails, whereas fish swim by lashing their vertical tails from side to side.

Dugong

Dugongs have horizontal tail-flukes like whales but inhabit only inshore waters.

The Dugong is an unmistakable Australian mammal that inhabits the sheltered coastal waters of northern Australia, mainly from about Shark Bay in the west to Moreton Bay in the east. It is also widespread from the shores of Africa to the Pacific. Totally aquatic, it never comes ashore. It prefers marine environments and only occasionally enters freshwater at the mouths of rivers. Feeding mainly on marine grasses, it also eats seaweed, especially when storms cause local destruction to the seagrass beds, and crabs and other crustaceans have also been found in their stomachs.

Adults are about 2–3 m long, occasionally larger, and may weigh 300 kg or more. There are no external hindlimbs and the relatively small forelimbs are used to 'walk' along the bottom of shallow waters only when Dugongs are grazing on seagrasses. When the animal is swimming rapidly it is the vertical strokes of the wide, lobed tail that provide propulsion.

SLOW BREEDERS

Having no competitors and few predators — except humans — the Dugong's reproductive rate is unusually low. Only at 4–8 years of age do they achieve sexual maturity and they bear only one calf at a time. The gestation period is about one year and youngsters remain dependent for up to two years.

Endangered Species

Although still reasonably common in parts of its Australian range, the Dugong is critically endangered almost everywhere else. Large and easily caught, it is avidly hunted for food by many local people. Its habitat and lifestyle also put it at constant risk of death from injuries caused by the propellers of pleasurecraft.

Sea-lions

The Australian Sea-lion is confined to coasts of southern and southwestern Australia, where it frequently hauls out onto beaches.

The coasts of southern Australia harbour several sea-lions and fur-seals but the Australian Sea-lion is the only one entirely confined to Australia. It occurs along the western and southern shores of the continent, mainly between the Houtman Abrolhos Islands and Kangaroo Island. It once occurred much further east but populations in Bass Strait were exterminated by sealers late in the nineteenth century. Rigorously protected, its populations now appear stable but it is estimated that total numbers remain something under 5,000 individuals.

Males are up to 300 kg in weight, nearly three times heavier than females. They are blackish brown, with a distinctive pale nape and crown; females are greyish above, creamy below.

Social Groups

The Australian Sea-lion is mainly sedentary and favourite beaches may be populated at any time of year. They are social animals, commonly gathering together in groups of 10–15 individuals of all ages and either sex but seldom containing more than one adult male. Most breeding colonies are on rocky shores but sandy beaches are often favoured for loafing ashore. Surprisingly agile on land, these marine creatures have been found several kilometres inland and can even climb low cliffs.

BREEDING SEASON

Most other seals and sea-lions have a distinct breeding season but Australian Sea-lions are unusual in that pups can be found at almost any time of year.

Fur-seals

Once more widespread, the Australian Fur-seal now has only nine breeding colonies, all in Bass Strait.

Several species of fur-seals occur throughout the islands of the Southern Ocean and from time to time strays appear on the shores of southern Australia but the New Zealand Fur-seal has substantial breeding colonies along Australia's southern coastline, mainly in South Australia and Western Australia. The considerably larger Australian Fur-seal, which also occurs in southern Africa, used to be quite widespread in southeastern Australia but current breeding colonies are confined to small islands in Bass Strait.

LIKE CLOCKWORK

Not all animals breed according to a seasonal or annual rhythm. Some populations of the Sooty Tern, a widespread tropical seabird, for example, nest to a schedule of nine months, and the Australian Sea-lion is unique among all pinnipeds in breeding on a cycle of 17.6 months. How such animals keep their 'internal clocks' coordinated with others of their kind is still not understood.

Australian Fur-seal

Like the Australian Sea-lion, the Australian Fur-seal is largely sedentary and lives mostly in coastal waters. Mainly warm brown in colour and around 300 kg in weight, males are much larger than females and have a distinct mane of long, coarse fur around their massive shoulders. Australian Fur-seals breed over a period of six weeks in November and December. Breeding sites often consist of densely packed herds of hundreds of females and immature fur-seals. Mature males defend territories against rival males and attempt to mate with any female that passes through their holding.

Dolphins

Like all marine mammals, Bottlenose Dolphins can dive deeply (above) but must surface to breathe (right).

Dolphins are familiar to many boat-owners as they are playful, sociable creatures that often follow a vessel's wake or ride its bow-wave, apparently just for fun. There are many species of dolphins but some are deep-ocean species seldom seen close to shore. The Bottlenose Dolphin is a conspicuous exception that is common in coastal waters all around Australian shores and frequently enters large sheltered harbours and bays.

Bottlenose Dolphin

This dolphin is easily identified by its plain, slate grey upper parts merging to pale grey flanks and white or pinkish white belly and underparts. It has a short beak with a distinct fold along its junction with the forehead, and the upswept jaw line gives it an inimitable 'grin'. Size varies with sex, age and locality but about 3–3.5 m in length and 150–200 kg in weight is typical.

Bottlenose Dolphins usually associate in groups or 'pods' of about 2–15 individuals; these may join other pods to form temporary aggregations of several hundreds. Their diet consists almost entirely of fish, which they often cooperate together in catching. Some, for example, may crowd a school of fish close into shore, while others patrol offshore to cut off their escape. Bottlenose Dolphins often escort yachts and pleasurecraft, gambolling in the bow-wave.

Humpback Whale

With their frequent 'breaching' and distinctive tail-flukes, Humpback Whales are easily identified.

Whales are notable for their spectacular 'breaching' behaviour in which they rear out of the water and fall back with a thunderous splash, often — apparently — just for fun. The humpback indulges in such activities with enormous exuberance. Adults are about 12–15 m long and are easily identified by their huge front flippers — nearly 4 m long. The tail is mainly white below but with variable black markings: these are a 'finger-printing' feature that scientists have long used to compile dossiers on the habits of individual humpbacks.

Distribution and Feeding

Humpbacks are widespread in the world's oceans but migrate annually from polar seas to breed in subtropical waters. They are especially easy to see in Australia because two of their most important migration pathways run close to both the east and west coasts of the continent and in some places they can even be seen from land.

The Humpback Whale is one of the 'gulp-feeding' whales that prey largely on krill — minute sea creatures that gather in enormous shoals. They capture them by taking huge mouthfuls of water and sieving the prey through comb-like curtains of baleen that hang from the upper jaw; with the water drained away, the food is left behind to be swallowed.

Right Whales

The Southern Right Whale lacks a dorsal fin and has distinctively rounded tail-flukes.

As the names suggest,there are two species of right whales: the Southern Right Whale is widely distributed in the oceans of the Southern Hemisphere and the Northern Right Whale is found in the Northern Hemisphere.The name 'right' dates back to the early days of whaling when the animals were harpooned from small boats: right whales were 'right' to attack because they swam slowly, did not sink when harpooned and have abundant stores of blubber. Unlike many other whales, right whales have no dorsal fin, and they also have distinctive, chalk-white callosities on the top of the head.These are distributed in a way that varies from one animal to another, enabling experienced researchers to identify them individually.

Southern Right Whales

Like the Humpback Whale, Southern Right Whales feed on krill.They migrate from summer feeding grounds in Antarctic waters to breed in warmer, temperate waters that include much of the coasts of southern Australia between about Perth and Sydney. In recent years the Great Australian Bight has been discovered to be a major nursery ground for these gentle giants of the deep, and mothers caring for their calves can be viewed from several vantage points along the cliffs.

RECORD-BREAKER

The largest of all animals is a whale. With a total length of about 28 m, and weighing up to 200 tonnes, the Blue Whale is not only the largest of all mammals — it is probably also the largest animal that has ever lived.

A Checklist of Common Australian Mammals

The following mammals are listed under their family names. Their scientific names further indicate their relationships to one another. Those species not listed here are either very rare or confined to remote parts of the continent.

ORNITHORHYNCHIDAE
Platypus *Ornithorhynchus anatinus*

TACHYGLOSSIDAE
Short-beaked Echidna *Tachyglossus aculeatus*

DASYURIDAE
Mulgara *Dasycercus cristicauda*
Little Red Kaluta *Dasykaluta rosamondae*
Kowari *Dasyyuroides byrnei*
Western Quoll *Dasyurus geoffroii*
Northern Quoll *D. hallucatus*
Spotted-tailed Quoll *D. maculatus*
Eastern Quoll *D. viverrinus*
Southern Dibbler *Parantechinus apicalis*
Northern Dibbler *P. bilarni*
Fat-tailed Pseudantechinus *Pseudantechinus macdonnellensis*
Woolley's Pseudantechinus *P. woolleyae*
Tasmanian Devil *Sarcophilus harrisii*
Fawn Antechinus *Antechinus bellus*

Yellow-footed Antechinus *A. flavipes*
Swamp Antechinus *A. minimus*
Brown Antechinus *A. stuartii*
Dusky Antechinus *A. swainsonii*
Brush-tailed Phascogale *Phascogale tapoatafa*
Common Planigale *Planigale maculata*
Wongai Ningaui *Ningaui ridei*
Southern Ningaui *N. yvonneae*
Kultarr *Antechinomys laniger*
Fat-tailed Dunnart *Sminthopsis crassicaudata*
Common Dunnart *S. murina*
White-tailed Dunnart *S. granulipes*
White-footed Dunnart *S. leucopus*
Long-tailed Dunnart *S. longicaudata*
Lesser Hairy-footed Dunnart *S. youngsoni*

MYRMECOBIIDAE
Numbat *Myrmecobius fasciatus*

PERAMELIDAE
Northern Brown Bandicoot *Isoodon macrourus*
Southern Brown Bandicoot *I. obesulus*
Long-nosed Bandicoot *Perameles nasuta*
Eastern Barred Bandicoot *P. gunnii*
Bilby *Macrotis lagotis*

PHASCOLARCTIDAE
Koala *Phascolarctos cinereus*

VOMBATIDAE
Common Wombat *Vombatus ursinus*

BURRAMYIDAE
Mountain Pygmy-possum *Burramys parvus*
Long-tailed Pygmy-possum *Cercartetus caudatus*
Little Pygmy-possum *Cercartetus lepidus*
Eastern Pygmy-possum *C. nanus*

PETAURIDAE

Striped Possum *Dactylopsila trivirgata*
Leadbeater's Possum *Gymnobelideus leadbeateri*
Yellow-bellied Glider *Petaurus australis*
Sugar Glider *P. breviceps*
Squirrel Glider *P. norfolcensis*

PSEUDOCHEIRIDAE

Lemuroid Ringtail Possum *Hemibelideus lemuroides*
Rock Ringtail Possum *Petropseudes dahli*
Herbert River Ringtail Possum *Pseudochirulus herbertensis*
Common Ringtail Possum *Pseudocheirus peregrinus*
Western Ringtail Possum *P. occidentalis*
Green Ringtail *Pseudochirops archeri*
Greater Glider *Petauroides volans*

TARSIPEDIDAE

Honey-possum *Tarsipes rostratus*

ACROBATIDAE

Feathertail Glider *Acrobates pygmaeus*

PHALANGERIDAE

Common Spotted Cuscus *Spilocuscus maculatus*
Common Brushtail Possum *Trichosurus vulpecula*
Scaly-tailed Possum *Wyulda squamicaudata*

POTOROIDAE

Musky Rat-kangaroo *Hypsiprymnodon moschatus*
Rufous Bettong *Aepyprymnus rufescens*
Brush-tailed Bettong *Bettongia penicillata*
Tasmanian Bettong *B. gaimardi*
Northern Bettong *B. tropica*
Long-nosed Potoroo *Potorous tridactylus*

MACROPODIDAE

Bennett's Tree-kangaroo *Dendrolagus bennettianus*
Lumholtz's Tree-kangaroo *D. lumholtzi*
Spectacled Hare-wallaby *Lagorchestes conspicillatus*
Agile Wallaby *Macropus agilis*
Antilopine Wallaroo *M. antilopinus*
Black Wallaroo *M. bernardus*
Black-striped Wallaby *M. dorsalis*
Tammar Wallaby *M. eugenii*
Western Grey Kangaroo *M. fuliginosus*
Eastern Grey Kangaroo *M. giganteus*
Parma Wallaby *M. parma*
Whiptail Wallaby *M. parryi*
Common Wallaroo *M. robustus*
Red-necked Wallaby *M. rufogriseus*
Red Kangaroo *M. rufus*
Bridled Nailtail Wallaby *Onychogalea fraenata*
Nabarlek *Peradorcas concinna*
Allied Rock-wallaby *Petrogale assimilis*
Herbert's Rock-wallaby *P. herberti*
Unadorned Rock-wallaby *P. inornata*
Black-footed Rock-wallaby *P. lateralis*
Proserpine Rock-wallaby *P. persephone*
Yellow-footed Rock-wallaby *P. xanthopus*
Red-necked Pademelon *Thylogale thetis*
Tasmanian Pademelon *T. billardierii*
Red-legged Pademelon *T. stigmatica*
Quokka *Setonix brachyurus*
Swamp Wallaby *Wallabia bicolor*
Banded Hare-wallaby *Lagostrophus fasciatus*

NOTORYCTIDAE
Marsupial Mole *Notoryctes typhlops*

PTEROPODIDAE
Common Blossom-bat *Syconycteris australis*
Eastern Tube-nosed Bat *Nyctimene robinsoni*
Bare-backed Fruitbat *Dobsonia moluccensis*
Spectacled Flying-fox *Pteropus conspicillatus*
Black Flying-fox *P. alecto*
Little Red Flying-fox *P. scapulatus*
Grey-headed Flying-fox *P. poliocephalus*

MEGADERMATIDAE
Ghost Bat *Macroderma gigas*

RHINOLOPHIDAE
Eastern Horseshoe-bat *Rhinolophus megaphyllus*

HIPPOSIDERIDAE
Dusky Leafnose-bat *Hipposideros ater*
Diadem Leafnose-bat *H. diadema*
Orange Leafnose-bat *Rhinonicteris aurantius*

EMBALLONURIDAE
Yellow-bellied Sheathtail-bat *Saccolaimus flaviventris*
Bare-rumped Sheathtail-bat *S. saccolaimus*
Common Sheathtail-bat *Taphozous georgianus*

MOLOSSIDAE
Northern Freetail-bat *Chaerephon jobensis*
Beccari's Freetail-bat *Mormopterus beccarii*
Eastern Freetail-bat *M. norfolkensis*
Southern Freetail-bat *M. planiceps*
White-striped Freetail-bat *Nyctinomus australis*

VESPERTILIONIDAE
Golden-tipped Bat *Kerivoula papuensis*
Little Bentwing-bat *Miniopterus australis*
Common Bentwing-bat *M. schreibersii*
Eastern Long-eared Bat *Nyctophilus bifax*
Lesser Long-eared Bat *N. geoffroyi*
Gould's Long-eared Bat *N. gouldi*
Greater Long-eared Bat *N. timoriensis*
Large-eared Pied Bat *Chalinolobus dwyeri*
Little Pied Bat *C. picatus*
Gould's Wattled-bat *C. gouldii*
Chocolate Wattled-bat *C. morio*
Hoary Wattled-bat *C. nigrogriseus*
Eastern False Pipistrelle *Falsistrellus tasmaniensis*
Western False Pipistrelle *F. mackenziei*
Large-footed Myiotus *Myotis adversus*
Greater Broad-nosed Bat *Scoteanax rueppellii*
Inland Broad-nosed Bat *Scotorepens balstoni*
Little Broad-nosed Bat *S. greyii*
Eastern Broad-nosed Bat *S. orion*
Western Cave Bat *Vespadelus caurinus*
Large Forest Bat *V. darlingtoni*
Eastern Forest Bat *V. pumilus*

Southern Forest Bat *V. regulus*
Eastern Cave Bat *V. troughtoni*
Little Forest Bat *V. vulturnus*

MURIDAE
Brush-tailed Tree-rat *Conilurus penicillatus*
Forrest's Mouse *Leggadina forresti*
Greater Stick-nest Rat *Leporillus conditor*
Broad-toothed Rat *Mastacomys fuscus*
Golden-backed Tree-rat *Mesembriomys macrurus*
Mitchell's Hopping-mouse *Notomys mitchelli*
Spinifex Hopping-mouse *N. alexis*
Fawn Hopping-mouse *N. cervinus*
Dusky Hopping-mouse *N. fuscus*
New Holland Mouse *Pseudomys novaehollandiae*
Silky Mouse *P. apodemoides*
Plains Rat *P. australis*
Western Pebble-mound Mouse *P. chapmani*
Desert Mouse *P. desertor*
Smoky Mouse *P. fumeus*
Long-tailed Mouse *P. higginsi*
Central Pebble-mound Mouse *P. johnsoni*
Hastings River Mouse *P. oralis*
Heath Rat *P. shortridgei*
Common Rock-rat *Zyzomys argurus*
Water-rat *Hydromys chrysogaster*
False Water-rat *Xeromys myoides*
Grassland Melomys *Melomys burtoni*
Giant White-tailed Rat *Uromys caudimaculatus*
Prehensile-tailed Rat *Pogonomys mollipilosus*
Long-haired Rat *Rattus villosissimus*
Bush Rat *R. fuscipes*
Swamp Rat *R. lutreolus*
Canefield Rat *R. sordidus*

DUGONGIDAE
Dugong *Dugong dugon*

OTARIIDAE
Australian Sea-lion *Neophoca cinerea*
New Zealand Fur-seal *Arctocephalus forsteri*
Australian Fur-seal *A. pusillus*

CANIDAE
Dingo *Canis dingo*

BALAENIDAE
Southern Right Whale *Eubalaena australis*

BALAENOPTERIDAE
Humpback Whale *Megaptera novaeangliae*

DELPHINIDAE
Common Dolphin *Delphinus delphis*
Bottlenose Dolphin *Tursiops truncatus*

INDEX

Antechinuses, 20
sex and ulcers, 23
Bandicoots, 24
Bats
biosonar abilities, 57
blindness, 56
Common Bentwing, 65
Eastern False Pipistrelle, 63
Eastern Horseshoe, 62
Eastern Tube-nosed, 62
Ghost, 59
Gould's Wattled, 65
Lesser Long-eared, 64
Little Forest, 64
megabats, 56
microbats, 60
observing, 61
pollinators, as, 61
Southern Freetail, 63
Bettongs, 40
Bilbies, 25
Brown Hare, 76
Brumbies, 77
Buffalo, Swamp, 79
Camel, One-humped, 78
Deer, Fallow, 74
Dingo, 74
Dolphins, 89
communication, 84
Dreys, 39
Dugong, 86
Dunnarts, 21
Echidnas, 11
brain-power, 13
hibernation, 13
European Rabbit, 77
Fallow Deer, 74
Feathertail Glider, 36
Feral
cat, 75
donkey, 79
goat, 77
pig, 78
Flying-foxes, 58
Fox, Red, 75
Greater Glider, 32
Hare, Brown, 76
Kangaroos
body temperature, 46
grey, 49
hopping, 46
Red, 48
suckling, 47
tree, 44
unusual reproduction strategy, 47
Koala, 26
diet, 28
finicky eating, 28
sleep, 28
Mammals
destruction, 80–81
diversity, 4
egg-laying, 9 –13
extinction, 80 –81
features, 5
gliding, 35
habitat, 4
introduced, 72 –81
marine, 82 –91
placental, 55 –71
tree-dwelling, 19, 34
types, 6
Marine mammals
fin-footed, 84
habitat, 84
Marsupials, 6, 15 –53
carnivorous, 19
export of, 38
predators, 22
tree-dwelling, 19, 34
Monotremes, 7
Mice
Mitchell's Hopping, 70
Western Pebble-mound, 71
Ningauis, 19
Numbat, 27
Pademelons, 50
Placentals, 6
Potoroos, 41
Platypus, 10
food location by, 12
spurs, 12
Possums
Brushtail, 37
communication, 38
evolution, 34
Green Ringtailed, 30
Honey-possum, 33
Leadbeater's, 31
Mountain Pygmy, 33
Striped, 30
Quokka, 43
Quolls, 17
Rabbit, European, 77
Rats
Broad-toothed, 68
Bush, 70
Common Rock, 71
Giant White-tailed, 69
Golden-backed Tree, 69
Water, 68
Red Fox, 75
Rodents
first wave of, 67
what are, 66
Sea-lions, 87
Seals, fur, 88
Spotted Cuscus, 32
Sugar Glider, 31
Swamp Buffalo, 79
Tails
prehensile, 38
tip, 18
Tasmanian Devil, 16
Tasmanian Tiger
extinction, 23
Wallabies
Hare, 42
Nailtail, 52
Rock, 53
Swamp, 51
Wallaroos, 45
Whales
beaching, 85
communication, 84
Humpback, 90
Right, 91
watching, 85
Wombats
subterranean lifestyle, 29